What Are the Chances?
Probability Made Clear

Professor Michael Starbird

THE TEACHING COMPANY ®

PUBLISHED BY:

THE TEACHING COMPANY
4151 Lafayette Center Drive, Suite 100
Chantilly, Virginia 20151-1232
1-800-TEACH-12
Fax—703-378-3819
www.teach12.com

ISBN 1-59803-267-4

Michael Starbird, Ph.D.

University Distinguished Teaching Professor of Mathematics,
The University of Texas at Austin

Michael Starbird is Professor of Mathematics and a University Distinguished Teaching Professor at The University of Texas at Austin. He received his B.A. degree from Pomona College in 1970 and his Ph.D. in mathematics from the University of Wisconsin, Madison, in 1974. That same year, he joined the faculty of the Department of Mathematics of The University of Texas at Austin, where he has stayed, except for leaves as a Visiting Member of the Institute for Advanced Study in Princeton, New Jersey; a Visiting Associate Professor at the University of California, San Diego; and a member of the technical staff at the Jet Propulsion Laboratory in Pasadena, California.

Professor Starbird served as Associate Dean in the College of Natural Sciences at The University of Texas at Austin from 1989 to 1997. He is a member of the Academy of Distinguished Teachers at UT. He has won many teaching awards, including the Mathematical Association of America's Deborah and Franklin Tepper Haimo Award for Distinguished College or University Teaching of Mathematics, which is awarded to three professors annually from among the 27,000 members of the MAA; a Minnie Stevens Piper Professorship, which is awarded each year to 10 professors from any subject at any college or university in the state of Texas; the inaugural award of the Dad's Association Centennial Teaching Fellowship; the Excellence Award from the Eyes of Texas, twice; the President's Associates Teaching Excellence Award; the Jean Holloway Award for Teaching Excellence, which is the oldest teaching award at UT and is presented to one professor each year; the Chad Oliver Plan II Teaching Award, which is student-selected and awarded each year to one professor in the Plan II liberal arts honors program; and the Friar Society Centennial Teaching Fellowship, which is awarded to one professor at UT annually and includes the largest monetary teaching prize given at UT. Also, in 1989, Professor Starbird was the Recreational Sports Super Racquets Champion.

The professor's mathematical research is in the field of topology. He recently served as a member-at-large of the Council of the American

Mathematical Society and on the national education committees of both the American Mathematical Society and the Mathematical Association of America.

Professor Starbird is interested in bringing authentic understanding of significant ideas in mathematics to people who are not necessarily mathematically oriented. He has developed and taught an acclaimed class that presents higher-level mathematics to liberal arts students. He wrote, with coauthor Edward B. Burger, *The Heart of Mathematics: An invitation to effective thinking*, which won a 2001 Robert W. Hamilton Book Award. Professors Burger and Starbird have also written a book that brings intriguing mathematical ideas to the public, entitled *Coincidences, Chaos, and All That Math Jazz: Making Light of Weighty Ideas*, published by W.W. Norton, 2005. Professor Starbird has produced three previous courses for The Teaching Company, *Change and Motion: Calculus Made Clear*; *Meaning from Data: Statistics Made Clear*; and with collaborator Edward Burger, *The Joy of Thinking: The Beauty and Power of Classical Mathematical Ideas*. Professor Starbird loves to see real people find the intrigue and fascination that mathematics can bring.

Acknowledgments

These lectures were prepared in collaboration with Thomas Starbird, Ph.D., a principal member of the technical staff at the Jet Propulsion Laboratory, Pasadena, California. Michael and Thomas Starbird were assisted by Nathanael Ringer, a Ph.D. student in Financial Mathematics at The University of Texas at Austin. Thanks to Lucinda Robb, Noreen Nelson, Pamela Greer, Alisha Reay, and others from The Teaching Company, not only for providing excellent professional work during the production of this series of lectures, but also for creating a supportive and enjoyable atmosphere in which to work. Finally, thanks to my wife, Roberta, and children, Talley and Bryn.

Table of Contents
What Are the Chances?
Probability Made Clear

What Are the Chances?
Probability Made Clear

Scope:

Many of the most significant events of our lives involve random chance—the people we meet, the accidents that befall us, the weather, the stock market, the games we play, the professions we fall into. Whether we are assessing the chance of being struck by lightning, the chance of winning the lottery, or the chance that it will rain tomorrow, we are confronted with trying to describe in as precise a manner as possible the likelihood of an outcome that is uncertain. Probability is the study that accomplishes the seemingly impossible feat of giving a meaningful numerical value to the likelihood that an event will occur when we admit that we do not and cannot know what will happen.

The basic strategy of probability is clear and simple. When we flip a fair coin, one of two equally likely outcomes will occur; namely, it will land on heads or tails. Thus, we define the probability of landing on heads as 1 out of 2, that is, 1/2. Or, if we roll a fair die, because there are six equally likely possible outcomes, the probability of rolling any one of them, say a four, is simply 1 out of 6, or 1/6. In other words, when probability involves equally likely outcomes, the concept of probability is simply a matter of counting. However, we soon find that the "simple matter of counting" is often not simple at all and frequently leads to surprises. A famous example is that among any random group of 50 people, there is a 97% chance that two or more of them have the same birthday. Our intuition about the likelihood of events, particularly rare events, often diverges sharply from the truth. As we explore probabilistic surprises, we will refine our intuition about the probability of random events and will learn more specifically what is surprising and what is not. We will learn why coincidences are so common and why we must learn to expect the unexpected.

In no place is the role of probability clearer than in games of chance; thus, we will introduce some of the basic ideas of probability using cards, dice, and roulette. In fact, it was in the arena of gambling that the mathematical investigation of probability first arose. In the 17th century, a gambler by the name of Antoine Gombault, the Chevalier de Méré, sought the advice of leading mathematicians of the day with the goal of improving his ability to make good decisions when playing dice. In answering Gombault's questions, Pierre Fermat and Blaise Pascal developed the fundamental concepts of probability.

Probability is the study of events whose outcomes are random. But randomness is a subtle concept. Events with random outcomes have the property that no particular outcome is known in advance; however, in the aggregate, the outcomes occur with a specific frequency. For example, when we flip a fair coin, we do not know how it will land, but if we flip the coin millions of times, we know that it will land heads up very close to 50% of the time. The distinction between our ignorance about the outcome of a particular trial and our knowing the aggregate behavior of many trials is the peculiar domain of randomness and probability.

Probability has applications in many arenas. For example, randomness and probability are central to the concept of statistical inference. But surprisingly, probability is involved in the solutions to many questions that do not at first appear to contain any element of randomness. For example, there are methods by which one can take a very large number, such as one with several hundred digits, and test whether or not it is prime using methods that involve probability. It is certainly not obvious how randomness and probability could possibly play a role in such a situation, because ultimately, the number is prime or it's not—there is no randomness involved. Another application of randomness and probability occurs in psychology. If we want to train our dogs to respond to a signal and keep responding longest, the best method may be to reward them randomly rather than on any fixed pattern. In this way, the dog always has the hope that the next reward is just one more good deed away. Of course, applying these insights to the treatment of people is most suggestive. Other examples in which randomness and probability arise occur in game theory, the study of strategic decision-making. In game theory, often the optimal strategy is one that involves intentionally including randomness. Optimal business strategies or sports strategies often are probabilistic in nature rather than deterministic. This feature complicates the question of how to judge whether we have adopted a good strategy. When probability is involved, even the very best strategy can have a poor outcome by chance alone.

Einstein's famous quotation, "God does not play dice with the universe," expressed his philosophical resistance to the probabilistic nature of quantum mechanics. Quantum mechanics asserts that subatomic particles are not best described as being in a certain place at a certain time but, instead, are better described with probability distributions, suggesting that an electron has some chance of being at any location in the universe at any moment. In fact, randomness and probability lie at the heart of many of the scientific descriptions of the physical and biological worlds. The basic idea of genetic inheritance is that the parents randomly contribute different genetic material to offspring, which then determines many features of the children. Evolution relies entirely on probabilistic occurrences. But we do not need to look to grand scientific theories to find examples of probability. We see probability in the newspaper every day when we read a weather report that says there is a 30% chance of rain. We'll see what that statement actually means.

Probability is a fascinating study that has many real-world applications. But one of the most intriguing aspects of all is that the basic meaning of probability in the real world is not clearly agreed upon by probabilists. In a rough sense, some view probability as measuring an individual's assessment or belief of the likelihood of a future event, while others view the probability of a future event as a fact independent of any individual's opinions. Another kind of distinction is that some probabilists allow probability to be applied to statements that do not entail randomness, such as "There was life on Mars," whereas others feel that probability should refer only to repeatable events with random outcomes. The different views of probability are intriguing to consider and, in some cases, have practical implications. Probability presents us with a rich field of intriguing inquiry that contains questions and insights that are mathematical, practical, and philosophical.

Lecture One
Our Random World—Probability Defined

Scope:

In many arenas, our understanding of our world involves processes and outcomes that we view as the result of random chance. We read in the newspaper that there is a 30% chance of rain. We talk about the chance of winning the lottery. Over the last century, scientific descriptions of the world have increasingly included probabilistic components. In quantum mechanics, the very location of subatomic particles is viewed as a matter of probability. The central concept of genetic inheritance and evolution is the random transmittal of genetic material from parents to offspring. Random happenings are those whose individual outcomes we do not or cannot know in advance but that will display regularity in the aggregate. The amazing accomplishment of probability is to put a meaningful numerical value on things we admit we do not know. Our challenges in this course are to understand what that numerical measure of chance is, to develop an intuition about probability in real-life situations, and to see a myriad of applications of probability in games, science, business, and many other aspects of life.

Outline

I. What are the chances?

    A. If you buy a lottery ticket, what are the chances that you will be rich?

    B. If you walk across a golf course on a stormy day, what are the chances that you'll be hit by lightning?

    C. If you bet on red in roulette, what are the chances you'll win?

    D. If you buy stocks and bonds, what are the chances those investments will pay off?

    E. If you have a fever and other symptoms, what are the chances you have a serious disease?

    F. A hurricane is spotted off the East Coast. What are the chances that it will cause great damage?

    G. What are the chances that a child brought up by a drug addict will become a criminal?

H. What are the chances that an e-mail advertisement will lead to a sale?

I. All these examples are real-life situations in which we are confronted with possibilities whose outcomes we do not know.

II. Dealing with the uncertain and the unknown is the realm of probability.

    A. One of life's challenges is to deal with the uncertain and unknown effectively.

    B. Probability accomplishes the amazing feat of giving a meaningful numerical description of the uncertain and unknown. It gives us information to act on.

    C. Probability decisions can be as inconsequential as deciding whether or not to take an umbrella if there is an 80% chance of rain.

    D. Making medical decisions based on probability, however, can have life-and-death consequences.

III. In many arenas, our understanding of our world involves processes and outcomes that we view as the result of random chance. Over the last two centuries, scientific descriptions of our world increasingly include probabilistic components.

    A. Physics, from thermodynamics to quantum mechanics, involves questions of probability—molecules moving randomly around and causing things to happen by the aggregate force of probabilistic occurrences.

    B. In biology, genetics and evolution are both based on random behavior.

    C. Often, underlying random behavior manifests itself in predictable, measurable observations. Scientific descriptions frequently are probabilistic analyses of random occurrences.

    D. The prevalence of probabilistic components of scientific descriptions represents a major paradigm shift in our concept of what scientific explanations are.

IV. Probability describes what we would expect from random phenomena if they were repeated many times. But the concept of randomness is subtle.

    A. Outcomes of individual random events are unknown, but the aggregate behavior of random events is predictable.

    B. The amazing accomplishment of probability is to put a meaningful numerical value on things we admit we do not know.

        1. When we roll a fair die, we do not know which side will land uppermost on any individual throw.

        2. However, if we roll 60 dice, we expect that each side would land up about $\frac{1}{6}$ of the time.

        3. One of the difficulties of probability is that we expect a certain result on average, but we also expect to be off by a little. When we roll 60 dice, we do not expect each number to appear exactly 10 times.

    C. One of the challenges of this course is to understand what to expect from randomness. A principal goal of probability is to give a numerical measure of chance.

    D. We will see a myriad of applications of probability in games, science, business, and many other parts of life.

V. The course is organized as follows:

    A. In this lecture, we will introduce the basic idea of probability.

    B. Lecture Two explores the question: What is randomness?

    C. Lecture Three is about expected value.

        1. Expected value is a numerical measure that assesses the value of various possible outcomes to a probabilistic occurrence.

        2. Expected value is useful in making decisions, such as those involving investments or other risks.

    D. Lecture Four takes us on a random walk, in which the direction we take at each step is randomly selected. Random walks have applications in physics, biology, and finance.

    E. Lectures Five and Six show us that randomness and probability are central components of modern scientific descriptions of our world in physics and biology.

F. Lecture Seven explores the world of finance, particularly probabilistic models of stock and option behavior.

G. Probability can be used to find answers to questions that seem to have no random or probabilistic component to them. Lecture Eight explores unexpected applications of probability.

H. Lectures Nine and Ten discuss conditional probability and some surprisingly counterintuitive examples of probabilities.

I. One view of probability is that it can describe a level of belief. Lecture Eleven explores this perspective and the Bayesian view of probability.

J. In the final lecture, we will see some probabilistic conundrums that arise when there are infinitely many possible outcomes to a random trial. We end by reviewing how widely probability is applied in the world.

VI. We begin our investigation of probability where it began historically, with gambling. Gambling presents some clear examples of randomness.

A. It was in the arena of gambling that the mathematical investigation of probability first arose. In the 17th century, a gambler by the name of Antoine Gombault, the Chevalier de Méré, sought the advice of Pierre de Fermat and Blaise Pascal, who developed the fundamental concepts of probability.

B. A die has six sides. In a fair die, we presume that after rolling the die, any one of the sides is as equally likely to arise as any other.

  1. To give a numerical measure to the probability of a fair die coming up with a five, say, we note that there are six equally likely possible outcomes; a five is one of these outcomes, so its probability of arising is 1 out of 6, or $\frac{1}{6}$.

  2. In general, if there are n equally likely outcomes, then the probability of one of those outcomes occurring is $\frac{1}{n}$.

C. Gambling games present us with examples in which there are finitely many possible outcomes to the probabilistic occurrence, that is, *discrete probability*.

VII. The concept of probability arising from dice and coin examples leads us to some basic definitions and observations about discrete probability.

A. An *outcome* is a possible result of a single trial, observation, or experiment that we are considering.

B. An *event* is a set of outcomes.

    1. For example, if we consider rolling a die, getting a five is an outcome.

    2. Rolling an even number is an event.

C. Probability 1 (or, equivalently, 100%) means that the event is certain.

D. Probability 0 means that the event will not happen.

E. If we add up all the probabilities of all the possible outcomes of a trial, we get 1.

F. If the probability of an event is p, then the probability of the event's not occurring is $1 - p$.

    1. For example, the probability of rolling a fair die and getting a five is $\dfrac{1}{6}$, so the probability of rolling a fair die and getting something other than five is $1 - \dfrac{1}{6} = \dfrac{5}{6}$.

    2. In practice, it is often easier to measure the probability that an event does not happen; for this reason, we will use the $1 - p$ observation frequently.

VIII. The basic principle of probability is simple when dealing with equally likely outcomes. Simply count how many total outcomes are possible, count how many are in the event you are considering, and divide.

A. The problem is that "simply" counting is not simple.

B. Let's think about poker.

C. The value of hands is really an ordering of the probabilities of getting the hands.

D. What is the probability of getting all four aces when dealt five cards?

1. To compute the probability of being dealt all four aces, we need to count the total number of five-card hands and compute the total number of hands that contain all four aces. Here are the answers:

2. The number of possible hands containing all four aces is $52 - 4$, or 48. The 4 represents the four aces, leaving only 48 cards that could be the fifth card in a five-card hand.

3. We can also calculate the number of possible hands: $52 \times 51 \times 50 \times 49 \times 48 = 311,875,200$.

4. But some of those hands will have the same cards, only in a different order; thus, we calculate the total number of different orderings of the five cards: $5 \times 4 \times 3 \times 2 \times 1 = 120$.

5. The number of distinct five-card hands is $\dfrac{52 \times 51 \times 50 \times 49 \times 48}{5 \times 4 \times 3 \times 2 \times 1} = \dfrac{311,875,200}{120} = 2,598,960$.

6. The probability of getting four aces is computed by dividing the total number of possible hands with four aces (48) by the total number of possible hands: $\dfrac{48}{2,598,960} = 0.00002$.

E. To compute the probability of being dealt a straight (see Glossary for definition), we need to count the total number of five-card hands and compute the total number of hands that contain a straight. Here are the answers:

1. The number of possible hands is 2,598,960.

2. The number of possible hands containing a straight is 10,200.

3. The probability of getting a straight is $\dfrac{10,200}{2,598,960} = 0.004$.

F. To compute the probability of being dealt a flush, we need to count the total number of five-card hands and compute the total number of hands in which all the cards are in the same suit. (Again, straight flushes are not counted as flushes.) Here are the answers:

1. The number of possible hands is 2,598,960.

2. The number of possible hands containing a flush is 5,108.

3. The probability of getting a flush is $\dfrac{5,108}{2,598,960} = 0.002$.

G. Because the probability of being dealt a flush is less than the probability of being dealt a straight, a flush beats a straight in poker.

IX. In summary, if you have an experiment or a trial that has equally likely outcomes, to compute the probability of some event, you count the number of outcomes in the event and divide by the total number of outcomes possible. That fraction is the probability of that event.

Readings:

Edward B. Burger and Michael Starbird, *The Heart of Mathematics: An invitation to effective thinking*, 2nd ed.

Edward B. Burger and Michael Starbird, *Coincidences, Chaos, and All That Math Jazz: Making Light of Weighty Ideas*.

Ian Hacking, *The Taming of Chance*.

Sheldon Ross, *A First Course in Probability*.

Questions to Consider:

1. Do you think that probability will play an increasing or decreasing role in explanations in science, business, social science, and other fields as they continue to develop?

2. Three couples, that is, six individuals, are seated randomly around a round table. What is the probability that the members of at least one couple are seated next to each other?

Lecture One—Transcript
Our Random World—Probability Defined

What are the chances? You buy a lottery ticket; what are the chances that you're going to be rich for the rest of your life?

You walk across a golf course in a stormy day; what are the chances you'll be hit by lightning?

You go to Las Vegas and you gamble your life's earnings by betting a red on roulette; what are the chances you're going to be rich?

You need to invest for your future and live happily in old age, so you buy certain stocks and bonds. What are the chances that those investments will allow you to live happily for the rest of your days?

You have a fever; you have a cough. What are the chances that it's a serious disease rather than something trivial?

A hurricane is spotted off the Gulf Coast and it's headed toward New Orleans; what's the chance that it will hit and cause damage?

What are the chances that a child who's brought up by a drug addict will become a criminal?

What are the chances that when you send out an e-mail advertisement to ten million people telling them to buy a certain thing that they'll actually buy any of it?

All these examples are real examples of life situations where we're confronted with possibilities whose outcomes we do not know. In fact, I would argue that many or most parts of our lives—and the world and trying to understand the world—involve situations where we don't know what's going to happen. They involve the uncertain and the unknown.

It would be nice to say, "Well, our challenge in life is to get rid of uncertainty and be in complete control of everything." That is not going to happen. One of life's real challenges is to deal with the uncertain and the unknown in some sort of an effective way; and that is the realm of probability. Probability accomplishes the really amazing feat of giving a meaningful numerical description of things that we admit we do not know, of the uncertain and the unknown. It gives us information that we actually can act on.

For example, when we hear there's an 80% chance of rain, what do we do? We take an umbrella. Now, of course, if the next day comes and it doesn't rain, what do we say? Well, we say, "Well, there was a 20% chance it wouldn't rain. That's okay." If it rains, we say, "Oh, yes, the prediction was right. There was an 80% chance of rain."

Probability is a rather subtle kind of a concept because it can come out one way or the other, and still a probabilistic prediction can be viewed as correct—but decisions made on probability have all sorts of ramifications. In the case of the rain, all we risk is getting wet. But in many areas of making decisions on the basis of probability, there are very serious consequences. When we make medical decisions, for example, we are making decisions that are based on probabilities, and yet they have extremely serious consequences, including life and death consequences.

In the early days of probability, one of the first examples where probability was out in the public—this was before probability was viewed as a commonplace thing as it is today—but back in the middle of the 1700s, between 1750 and 1770 in Paris, there was a smallpox epidemic, and they had developed a smallpox vaccine, but the inoculations were rather risky. They reckoned that there was a 1 in 200 chance of death from taking the inoculation, but on the other hand, there was a 1 in 7 chance of dying eventually from the disease. So making that kind of decision is a very dramatic question where we're weighing probabilities. If you took that inoculation and you died immediately from smallpox, did you make the right decision or not? Well, of course, you don't want to be among the 1 in 200 that died from the inoculation. On the other hand, on the basis of probability, it was the right decision. There are many controversies about this kind of thing and in today's world with lawsuits and all this would be a very serious kind of an issue to undertake.

Well, in many arenas of life, our understanding of the world comes down to understanding processes and outcomes that are probabilistic in nature, that really come about from random chance, that things are happening by randomness alone. Over the last couple of centuries, the scientific descriptions of our world increasingly have included probabilistic components in them.

For example, in physics, many aspects of physics, from quantum mechanics to thermodynamics, all of them involve questions of probability. Things we imagine—molecules moving randomly around and causing things to happen by the aggregate force of probabilistic occurrences—quantum mechanics, the same thing—at the very foundations of our knowledge of physics is probability.

Biology, genetics, and evolution are all based very centrally on random behavior. In fact, in all of these areas, the goal is to make definite, predictable, measurable statements about what's going to happen that are the result of describing random behavior. In fact, the description of these kinds of probabilistic analyses of random behavior is actually what is involved in scientific descriptions of the world. This is a major paradigm shift in the way science has worked recently—by recently, I mean in the last 150 years—is that there has been an increase in the role of probabilistic and random descriptions as the center of scientific descriptions of the world.

One thing that probability tries to do is to describe random phenomena. It tries to give a specific statement about what we expect when things happen at random. The reason that it can be effective at doing this is that random happenings are things where the individual outcomes of one trial or one experiment are completely unknown, but if you repeat them many, many times, or you look at them in the aggregate, they have some regularity to them; and so the amazing accomplishment of probability is to put a meaningful numerical number value on the things that we admit we don't know.

I want to give you an example that I think captures this whole idea very clearly. This is a die. It has six sides, and if I roll this die, it has an equally likely chance to come out any of one of the numbers 1 through 6. So I'll just go ahead and roll it, and I don't know what's going to happen. Okay, it came out a 2—could have come out 1 through 6. I don't know anything about what could have happened. But let me do a similar experiment, except instead of with just one die, I'm going to use 60 dice. So here I have an urn—by the way, in probability we always talk about picking things from urns—so this is now an urn for your information and in it there are 60 dice. Now, I'm going to take these 60 dice, and I'm going to shake them up and roll all of them.

Now, when I rolled one die, I didn't know what was going to happen, but I claim, without even looking at this, I'm going to show you that just by randomness alone, I have an expectation that there'll be approximately as many 1s as 2s as 3s as 4s as 5s and as 6s, even though I didn't do anything. You saw me. I just shook them up and rolled them. And so I'm going to take a second here and arrange them neatly so we can count how many 1s, 2s, 3s, 4s, 5s, and 6s there are.

Okay, we've now arranged the dice so that we can see exactly how many there are of each of the different numbers. Now, in this case actually there's more variation than I usually get in this, but I think maybe we should go along with it to demonstrate that there is variation to expect also. We would not expect each of the numbers to come out exactly 10. That would be a very unusual occurrence. Instead, what we expect is that on average, of course, on average there'll certainly be 10, because there are 60 dice and 6 possible numbers to come up; but notice, for example, there was no number that didn't appear at all. The least number here, the 4s, there were six 4s; and the most number, the 3s, there were fifteen 3s. So there was certainly some variety around the number 10, but nevertheless, our expectation was that on average that we would expect a rather even distribution of these different numbers 1 through 6. We would not expect all of them to come out 6s, and as I say, this is a more extreme variation from the numbers than I usually get when I do this experiment; but that happens, too. With probability, you get these kinds of variations.

The role of probability is to try to describe the operation of random occurrences in the aggregate. One of the challenges for this course is to understand what to expect from randomness, and that is the role of probability: to describe what to expect from random phenomena. The goal of probability is to give a numerical measure of those chances. We're going to then see a myriad of applications, by the way, of probability in all sorts of things, from games, to science, to business, and many other parts of life.

Let me take just a few minutes to tell you how the whole course will be organized. In this lecture—besides this introduction—we're going to introduce the basic ideas of probability. And then in the next lecture we're going to talk about the question of "What is the nature of randomness?" And this is going to be the lecture in which we're going to see some of the surprising issues that are associated with randomness.

In Lecture Three we're going to introduce the concept of expected value. Expected value is a measure, using which we can make decisions about probabilistic kinds of outcomes. It gives us a numerical way to say whether or not to bet on this or that game or make this kind of an investment.

In Lecture Four, we're going to take a random walk. A random walk is a description of random fluctuations, and we'll see these appear in all sorts of application areas, such as physics, and biology, and finance.

In Lectures Five and Six, we're going to see how probability plays a role, first in the physical sciences, and then in Lecture Six in the biological sciences, in modern scientific descriptions of these scientific areas.

And then in Lecture Seven, we're going to see how the world of finance is described using probabilistic models, including models of stocks and option behaviors.

In Lecture Eight, we're going to talk about how probability appears where we don't expect it; so finding probability where we don't expect it, and there are all sorts of amazing areas that seem to have no random component to them at all, and yet where probability can play a central role.

In Lectures Nine and Ten, we're going to discuss conditional probability and give some of these really famous examples that any course in probability must have, and they're just wonderfully interesting things to discuss, such as the birthday problem, for those of you who know that, but you can hear that in Lectures Nine and Ten.

And then in Lecture Eleven, we're going to look at probability from a different point of view: namely, probability associated with measuring a level of belief, instead of a measure of the frequency with which a random process occurs. This is the Bayesian view of probability.

And then in the final lecture, we're going to see some examples that challenge our intuition about probability, and then show, once again, how prevalent probability is in our world.

But we're going to start today to introduce the very basics of probability by introducing it where the investigation of probability began historically, mainly with gambling. Gambling presents us with all sorts of clear examples of randomness and of probabilistic analysis, and so that's where it started. That's where the mathematical investigation of probability first arose. It began in the 17th century when a gambler by the name of Antoine Gombault, the Chevalier de Méré, sought the advice of two famous mathematicians of the day, Pierre Fermat and Blaise Pascal, and those two—Fermat and Pascal—developed the fundamentals of probability, although it had been discussed somewhat before.

So let's begin with taking a die, and explaining the most basic concept of probability. Remember the concept is that we wish to associate a number, giving a numerical value, to a random process—to a process where there are several different possible outcomes, but we don't know which one will happen. In the case of a die, there are six sides to a die, of course, and consequently we would say that the probability of any one of those arising is as equally likely to happen as any of the other sides arising if we rolled the die, and consequently we give a number to that by saying there is a 1 out of 6 chance that, for example, a 5 will arise if I roll a die. And consequently that is the very most basic concept of probability—that if we have equally likely possible outcomes of some experiment or some trial, then we'll say that we just count how many outcomes there are that are equally likely and then the chance of any one of them occurring is 1 over the number of outcomes that are possible. So in the case of a die, the probability of coming up a 4 is 1 out of 6.

Okay. Now, by the way, what we're going to be discussing in the first part of these lectures is a field called discrete probability, where we're just talking about a finite number of the possible outcomes that can happen, and that will start to simplify our discussion, and then later, we'll extend that to other realms. In the case of a die, there are six outcomes that are equally likely to occur. If I take another paradigm of probability, namely a coin, and I flip the coin, there are two equally likely outcomes that could occur—a head or a tail.

Well, those possible outcomes are called *outcomes*, and if we have a collection of outcomes, such as, for example, asking the question: What's the probability of rolling a die and getting an even number? then that's called an *event*. The collection of outcomes that are even is called an event, and I could ask: What is the probability of rolling an even number? And of course to compute the probability of rolling a die and getting an even number, it's very simple. It's just—since there are three even numbers, 2, 4, 6—the probability is just 3 out of 6.

These are the very most basic concepts of probability. To summarize, let me just point out that probability has a couple features: One is that the probability of a particular outcome or of an event is equal to a number, and that number is always a number between 0 and 1. The number 1 means that the event is definitely going to happen. It would be like, "What's the probability that I roll a die and I get some number?" Well, it's going to happen. That's probability 1; and probability 0 means that the event will not happen. If I have just finitely many possibilities, and I roll a die, and I say the event is that nothing will come up, well, that's not going to happen—probability 0.

If you look at all of the outcomes of a particular trial or experiment, and this could happen, or that could happen, or the other thing could happen, the probabilities—the sum of those probabilities—of the different possible outcomes, if you list all of them, the sum has to be equal to 1, because the probability that something will happen is 1. These are just some very basic things.

The second one is if you have the probability that an event has happened; for example, the probability that I roll a 5 if I roll a die, is 1 out of 6, so the probability of the event of not rolling a 5—that is, everything else—is going to be $(1 - 1/6)$ or $5/6$.

These are some basic things, and as we come to them in actually computing things, we'll refer to them again. But these are some basics of probability, and I would say that this next equation is one that summarizes the basics of probability most clearly, and that is: That if we have a collection of outcomes that are equally likely, and we want to compute the probability of a particular event (E)—and remember an event is just some set of those outcomes—then the probability is just a fraction, and the denominator of the fraction is the total number of outcomes, and the numerator of the fraction is the number of outcomes that are in the event.

As I said before, if I want to say, "What's the probability of rolling an even number?" Well, there are three even numbers, so that's the numerator; and then there are six possible outcomes—1 through 6—so the probability of rolling an even number is 3 out of 6. Now, so what that tells us is that probability for cases where every outcome is equally likely to occur, the question of probability is a very simple question, because it's just a question of counting. You count how many things you're talking about in the event you're trying to describe, and divide by the total number of outcomes that are possible—very simple.

The only problem is that counting is anything but simple. It turns out that counting is one of the hardest things that we can do, and let me just demonstrate this by talking about an example that is very common to think about in probability: namely, talking about cards. Here I have a deck of cards. Now, if we deal out a poker hand, a poker hand consists of five cards. And so we can ask questions about the probability of getting various poker hands. So let's go ahead and be specific here, and think about a particular poker hand that we would like to get and see if we can compute the probability of getting it.

What's the probability of getting a poker hand that has all four aces? So, now, this is what you'd like, if you were sitting at a poker table and somebody dealt you the following hand, you'd be very happy because there are four aces there. Well, remember, the concept of probability and computing the probability is to count the number of outcomes that we're interested in—namely, those hands that have four aces—and find out how many there are among all possible ways of being dealt five cards. All possible ways of being dealt five cards, how many of them will contain four aces? And then all we do is we divide by the total number of five-card hands you could be dealt. Well, so all we need to do is count two things.

Well, let's go ahead and start by counting the easier one first. The easier one to count is the number of hands that contain four aces. The reason that this is an easier thing to count is because if four of the five cards in a randomly dealt hand are aces, there's only one card remaining, but that one card could be any one of the remaining cards in the deck. It could the queen of hearts; it could be the four of clubs; it could be the six of hearts; it could be the seven of spades; it could be *any* of the remaining cards in the deck. How many remaining cards are there in a deck of 52 cards if four of the cards in the hand are already aces? Well, the answer is there are only 48 remaining cards. That means among all the possible ways of getting five cards, there are only 48 ways of getting five cards that contain four aces; it just depends on that last card.

Okay. Half of our counting is done. We've counted how many outcomes have four aces in them—total 48. Now all we need to do is talk about the denominator of our fraction. Remember, to compute the probability we want to compute the number of hands that contain four aces divided by the total number of five-card hands that could be dealt. Well, how in the world are we going to count how many possible five-card hands there are total? In a deck of cards that you were dealt five cards, how would you compute such a thing? This is a difficult question.

Let's go ahead and figure out how to do it, and the way we do it is we analyze it in the following way. We say to ourselves, "Okay, let's imagine that we're being dealt these cards one at a time, so the first card that we're dealt is something—in this case it's the five of clubs." So we're dealt the five of clubs, and then we say to ourselves, "Aha, after being dealt the five of clubs, how many cards could appear as the second card in the hand that we're being dealt?" Well, it could be any one of the 51 remaining cards in the deck. So any one of these 51 different cards could be the second card that we're dealt. Now, for any one of those choices—for example, this two of hearts here choice—for any one of these choices of these being the first two cards, how many choices are there for the third card? Well, let's see, it could be anything that we can imagine—like it could be the king of diamonds, six of spades, it could be any of these things—how many are left to be chosen from? Well, if you're already picked two cards, there are 50 remaining cards.

Here in this graphic we can see this bifurcating tree of possibilities. For every first card there are 51 possible second cards. After choosing the first two cards, there are 50 possible third cards, there are 49 possible fourth cards, and there are 48 possible fifth cards possibly dealt to you as a hand.

Now, are we correct, then, in saying that the total number of five-card hands we could be dealt, which is five, or the number of cards in a standard poker hand, $(52 \times 51 \times 50 \times 49 \times 48)$—is that the correct answer for the number of hands that we could be dealt? By the way, that number, if we actually multiply it out, is 311,875,200. There are a lot of different ways to get five cards in order. But is this a correct count?

Counting is tricky, and the reason that it's tricky is because this would not be a correct count for the number of five-card hands you can get. And the reason is that—look at this hand that we've just counted, this was one of the hands we counted—the five of clubs, two of hearts, eight of diamonds, king of diamonds, seven of clubs. That was one of these three-hundred-eleven-million-something hands that we counted. But, notice that if we just change the order, and we put the two of hearts first and then the five of clubs and these other three cards, we would have counted that as a different way of getting five cards, right? Because this one started with the two of hearts, so it's different from the one that we counted starting with the five of clubs. But, nevertheless, this is the same poker hand. When you have a hand of cards you can move them around, it doesn't matter. And consequently we've counted this same hand—these same cards— many, many times. We counted this ordering as a different thing from this ordering as a different thing from this ordering as a different thing from this ordering. Every possible ordering has been counted separately, and yet it's the same hand.

What we need to do is think about how to systematically count how many times we have over-counted, and the answer is that we have over-counted by the number of ways we can take the same five cards and reorder them. Well, we could put any of the cards first, any of the four remaining cards second, any of the three remaining cards third, any of the two remaining cards last, and then the last card is determined. So we have a total of $(5 \times 4 \times 3 \times 2 \times 1)$ different ways of counting the same five cards that we have over-counted by. So to get the actual number of hands possible in poker, it's 2,598,960, because we take the 311,875,200 of hands that count the order in which we get them and divide by the fact that each hand has been counted 120 different times.

Now this, by the way, is the area of *combinatorics*; how to carefully count such things as this is a whole area of mathematics called combinatorics. Having explained all this, I hope that you forget it, because I want you to focus on the bottom line of what we're talking about: namely, that to compute probability we need to count the total number of hands with, in this case, four aces, and divide by the total number of hands to get the probability. The thing to keep your eye on is that probability is just the total number in the event divided by the total number of possible outcomes.

Let's talk about what's the probability of getting a straight. A straight is where the cards are in order—like 3, 4, 5, 6, 7—but they can be in any suit. Well, once again it's a matter of counting. By careful counting, and it's tricky. You can count that there are 10,200 poker hands that are straights, so the probability of getting a straight is that 10,200 divided by 2,598,960, which we already computed was the total number of poker hands that you can get, for a probability of .004.

We can do the same thing to compute the probability of getting a flush. A flush is a hand where all of five cards are in the same suit. The number of poker hands that are a flush are 5,108, it turns out by careful counting, so the probability of getting a hand, if just being dealt five cards, where all of them are the same suit, is only .002; that is, 2 out of 1,000. And by the way, there's a technicality for both straights and flushes. It's that something that's both a straight and a flush is not counted as a straight, nor is it counted as a flush; it's a different category. That's why if you try to do this at home you'll probably get a slightly different answer.

The point is that we've just computed that the probability of being dealt five cards that are a straight is 4 out of 1,000, and the probability of getting five cards where all the cards are in the same suit is only 2 out of 1,000. The way poker works is that the rarer the hand the higher the value, so a flush beats a straight at poker.

Okay, so to conclude this lecture then, I want you to focus on one thing: Namely, that the basics of probability are that if you have an experiment or a trial, such as being dealt five cards from a deck of cards, or rolling a die, if you're doing a trial which has equally likely outcomes, then if you want to compute the probability of some event, all you do is you count the number of outcomes in the event—such as the number of straights—and divide by the total number of outcomes that there are—such as the number of hands that could possibly be dealt—and that fraction is the probability of that event.

In the next lecture, what we're going to do is to take up the challenge of trying to understand the nature of randomness itself. I look forward to seeing you then.

Lecture Two
The Nature of Randomness

Scope:

What is random? Can we ascertain whether phenomena in the world are best described by randomness or are better described by finding some underlying deterministic reason for what we observe? Questions about what is random arise in considerations of everything from a coin toss to dots on a page, stars in the sky, or the digits of π. Trying to produce lists of numbers that appear random is an unexpected challenge. If we look at a list of digits, can we determine whether or not they were generated by a random process? Many tests about randomness can ferret out the signature of nonrandom generation. One of the paradoxes of randomness is that within the random, we will find surprising instances of patterns that occur by chance alone.

Outline

I. One goal of probability is to describe what to expect from randomness.

    A. The challenge is to understand in some detail the nature of random processes.

        1. Surprisingly, clear order comes from random activities.

        2. Randomness refers to situations in which we don't know any individual result, but we have a sense of what will happen in the aggregate, that is, if an experiment or a trial is done over and over again.

        3. This idea is captured in a theorem called the *Law of Large Numbers*.

    B. We can illustrate this theorem ourselves by doing various experiments, such as rolling a die and calculating the percentage of times we roll a three.

        1. The more times we roll the die, the closer we come to the predicted probability of rolling a three, $\frac{1}{6}$, or 0.1667.

2. Throwing the die 6 times, we might get no threes, but in rolling the die 60,000 times, we come very close to the expected 0.1667.

II. The Law of Large Numbers works even when referring to relatively rare events.

    A. If we draw one card at random from each of three decks, the probability that the three cards will be identical is quite small: $\dfrac{1}{52} \times \dfrac{1}{52} = \dfrac{1}{2,704}$, or 0.00037.

    B. After 2,704 trials, we got no such matches, but after 2,704,000 trials, we got 1,037 such matches: $\dfrac{1,037}{2,704,000} = 0.00038$, very close to the probability.

III. There are counterintuitive aspects of what is produced by randomness.

    A. A visual example illustrates this phenomenon: Working with a square, we pick a place on the vertical axis at random and on the horizontal axis at random and put a dot there. We do this 12 times to produce 12 dots.

        1. We expect the dots to be more evenly distributed rather than the clusters and gaps we see.

        2. We can also see other patterns in random arrays. Look at the night sky, for example, and see the various constellations that have been identified for centuries.

    B. Flipping a coin also illustrates randomness.

        1. First, we flip a coin and record the results, heads (Hs) and tails (Ts), over 200 flips.

        2. Then, we ask a human being to write down a random list of 200 Hs and Ts.

        3. Strings of repeated Hs or Ts in the flips show up more often than in the human-generated list of HTs.

        4. Specifically, when you flip a coin 200 times, the probability of having a string of six Hs or six Ts is more than 96% and of having a string of five Hs or Ts is 99.9%.

        5. Our simulation shows that even if you have flipped 10 Hs in a row, the next flip is just as likely to be H again as it was the first time you flipped the coin. The coin has no memory.

IV. Rare events are expected in probability.

    A. As we have seen, the probability of getting any particular five-card hand from a deck of cards, whether an ordinary hand or a royal flush, is $\dfrac{1}{2,598,960}$.

    B. The probability of winning the Powerball lottery is $\dfrac{1}{146,000,000}$, but someone is very likely to win.

V. Even very rare events are almost certain to happen given enough opportunities.

    A. In 1929, the astronomer Sir Arthur Eddington wrote, "If an army of monkeys were strumming on typewriters, they might write all the books in the British Museum." It is said, then, that if monkeys randomly type, they will eventually write *Hamlet*.

    B. Let's look at this further. If, since the time of the Big Bang, a billion 18-character patterns were generated per second on a 100-key keyboard, chances are less than $\dfrac{1}{1,000,000,000}$ that "To be or not to be" will be generated.

    C. An enterprising author made money with an observation a few years ago when he wrote *The Bible Code*. For example, he found that if he looked at every $1,945^{th}$ letter somewhere in the Bible, it spelled out "Atomic holocaust, Japan, 1945."

    D. Mathematicians found *mail* and *bomb* in Ted Kaczynski's manifesto.

VI. When we look retrospectively, things that appeared to be random can be explained.

    A. Stock movements can be explained in retrospect.

    B. Some psychics and stock analysts make correct predictions by chance alone.

VII. How can we distinguish a set that was created from a random process versus some other method?

A. The strategy is to analyze what patterns we would expect to occur by random chance.

B. Suppose we consider flipping a coin.

    1. Roughly half the results should be Hs and half Ts.

    2. As we flip more coins, that fraction should get closer and closer to 50%.

    3. We can get more refined and determine what fraction of HHs or TTs we should expect and so forth.

    4. We can compute the probability of each pattern. By seeing whether the appropriate frequency of that pattern appears or does not appear, we gain evidence about the likelihood that the list of Hs and Ts was generated randomly.

VIII. Some examples bring up challenging philosophical questions about the meaning of randomness.

    A. Consider the first 10,000 digits of π.

    B. The digits look random from the point of view of the tests concerning the existence of patterns, yet we know they are completely determined.

Digit	Number of Appearances in the First 10,000 Digits of π
0	968
1	1,026
2	1,021
3	974
4	1,012
5	1,046
6	1,021
7	970
8	948
9	1,014

    C. What kinds of events are actually random in the world and which are deterministic? These are issues that present us with a real philosophical challenge.

Readings:

Ivars Peterson, *The Jungles of Randomness: A Mathematical Safari.*

Questions to Consider:

1. Do you think that analyzing or modeling some phenomenon as if it were random devalues or depersonalizes the situation? Do you think that such an analysis skirts the actual meaning?

2. On learning that some girl in the neighborhood has committed a minor crime, how do you react to a statement such as: "Well, it was bound to happen. Statistics show that about 20% of kids do that."

Lecture Two—Transcript
The Nature of Randomness

Welcome back. The basic goal of probability is to describe what it is that we should expect from randomness, and so in this lecture we're going to try to undertake an understanding in some detail of the nature of random processes.

The first thing about randomness is that clear order actually arises from random processes, particularly if we do them enough times. Randomness refers to situations where we don't know what any individual result will be, but we do have a sense of what will happen in the aggregate or, that is to say, if the experiment or the trial is repeated over and over again.

This is captured in a theorem called the law of large numbers, which says the following thing: That if we conduct an experiment—and think of it as rolling a die or dealing cards—and there are many different possible outcomes that there could be, and we're inquiring about the probability of a particular outcome or event, then it turns out that if we do this over and over again, the fraction of the trials that exhibit that event outcome that we're looking for, divided by the total number of trials that we try, will get closer and closer to the probability of that event.

Now this is really a concept that is what you would guess intuitively. There's no surprise in the law of large numbers except that maybe that it actually works. Let's go ahead and look at some examples here. These were done by simulations of rolling a die. Now, suppose that we roll a die and we ask ourselves: What's the probability of getting a 3? Well, there, I got a 3 that first time. Now, what we did is we simulated on the computer rolling a die many, many, many times to see how many 3s we got, and to try to demonstrate this law of large numbers, which tells us that the fraction of 3s that we get over many, many repeated trials, that fraction, divided by the total number of trials we have—the number of 3s divided by the total number of trials—will become increasingly close to the probability. Now, we know what the probability is of getting a 3—it's 1 out of 6, or 0.1667—6666666 forever, actually, which we round to 0.1667.

We did this experiment several times of rolling the die: 6 times where we got no 3s in this experiment; 60 times where—you see, we expect on average to get about 10 out of 60—in fact, we got 9, and its ratio then was 0.15 (that is, $9 \div 0$ is 0.15). As we repeated the experiment more times—600 rolls, we got 92, for a fraction of 0.1533; 6,000 times, we expected 1,000, in fact got 997, very close: 0.1662; 60,000, we expected to get 10,000 3s; in fact, we got 10,037 3s, and so we got 0.1673. You can see that this last number here, after we repeated the experiment 60,000 times, that the fraction of times we get a 3 is getting very close to the actual probability.

We repeated the same experiment of rolling the die 60,000 times, and you can see in each case down here, the fraction that we get of the actual outcome approached very close to the expected probability of 0.1667. That is the law of large numbers in effect.

But let me tell you that the law of large numbers works even when we're talking about relatively rare events. And so let me explain that by giving you an example of a relatively rare event and seeing how the law of large numbers will apply. Suppose that we take three decks of cards, okay, and now we just pick a card randomly from each of these three decks. Well, we can ask ourselves the question: What is the probability if we pick one card from each deck that we will pick precisely the same card at random from each of those three decks? Well, it's really quite rare, quite rare, and we can be specific about how rare it is; namely, let's do an analysis to see what the probability is of picking three cards that are the same.

Well, when we pick a card from the first deck it doesn't matter what it is, because we have to match it by the card in the second deck. What's the probability that the card from the second deck will match? Well, it's 1 out of 52. And what's the probability that the card in the third deck will match that same thing? One out of 52 again. The probability of all of them being the same, which they weren't of course in this case, but the probability of their being the same is just 1 out of 52—that's the probability that the second deck will give a card that matches the first—times, of the times that the second deck matches, only 1 of 52 times will the card from the third deck also match. If the probability is the product of 1 out of (52×1) out of 52, which is 1 out of 2,704, or 0.00037; that is, 37 out of 100,000 times we would expect to get a match. Quite rare, quite rare.

Well, we actually undertook some simulations to see if we could actually see that the law of large numbers works even for very unlikely things, like picking three cards that are exactly the same. Here we go. We did a simulation of—first we drew just 2,704 times—we simulated drawing these cards, and none of them produced a match. By the way, we picked 2,704 because the probability tells us that 1 out of 2,704 times we should get the same. So then we tried 27,000 [sic 27,040]; we got 12. Our expected number was 10, and you can see that the ratio of the number of successes—that is, all three the same—divided by the number of trials—in this case 27,040—12 divided by 27,040 is 0.00044, and the actual probability, as we see in this column repeated, is 0.00037.

Well, when we did it 270,400 times, we got a ratio of 0.00030—getting close. When we did it more than 2,704,000 times, we expected to get about 1,000; in fact, we got 1,037 times that they matched exactly, which had a ratio of 0.00038, which you see just differs in that last decimal point from the actual probability that is expected. The law of large numbers is in operation even when the event is rather rare, such as getting three cards that are exactly the same.

This lecture is about the unexpected reality of what to expect from randomness, and there are counter-intuitive aspects of what is produced by randomness, and I love these. I love the idea that randomness produces things that are sort of surprising to us.

The next one I'm going to give is a visual example. Suppose that we undertake the following experiment: We just take a square, and in the square we pick some dots at random; that is, by picking them at random, I mean that we pick a place on the horizontal axis at random and the vertical access at random—we just pick a number—and then put a dot there. And we do that, in this case we did it 12 times, and drew 12 dots. What we did is we undertook that experiment of doing it several times, and so let me just show you what we got from just randomly picking 12 dots in a square.

Now look at all these things. Look at that. Look at that. The thing that a lot of people find surprising about these 12 random dots in a square is that they expect something that's a little bit different; these were all produced just by random process, of randomly choosing the location of the dots. Look at that. Isn't that surprising? See, look at that—that's a nice—like a question mark pattern.

In contrast, here are some non-randomly positioned collections of dots on the page. Now, the reason I draw these is that if I ask you to take out a piece of paper and randomly draw 12 dots on the page, I think you'd more apt to be drawing something like this. You'd be more apt to draw the dots so that they sort of evenly are spread out over the whole square than was the case in the actual randomly collected dots, and that's what is surprising about randomness, that it produces the unexpected—these big gaps in the clusters and things that we saw in the randomly generated positions.

And, by the way, in these randomly generated positions we can find all sorts of patterns that seem to just come by a miracle, and one place to look at this is in the night sky. When we look at the night sky, the night sky is sort of a random collection of dots, and so that's sort of a fun thing to look at. If we look at the night sky—of course, by the way, it's not random when we look at the Milky Way, because the Milky Way, it's not evenly likely to see a star that's out of the plane of the Milky Way—but if we look in some other direction, there's a rather random collection of dots, and sure enough, from ancient times people have found patterns among these stars that are randomly put there; namely, all the constellations. This is an example of the way in which we see patterns in randomness.

Let's move on to another example. We've talked about a visual example of randomness, the dots on a page. Let's talk about flipping a coin. If we take a coin and we flip it, we flip it many times, and we record whether it's a heads or a tails, and then we flip it again, we see whether it's heads or tails, and we record the flips that we make. Let's suppose that we do this 200 times, and we just write down what we get, and here is what it is, you don't have to look too carefully at this, but this is the actual result of flipping a coin 200 times at random, and actually it was simulated, and then putting down H's for the heads and T's for the tails.

Now, suppose that I ask a human being to write down a random list of 200 H's and T's, and just write it down—and I often do this when I'm giving talks to people, to say this, I say, "Okay, I want one group to actually flip a coin, and another group to just write down what you think is a random sequence of H's and T's—write anything down." Then I have them put it on the blackboard or display it, and I come in and I'll tell them which one was done by a human and which one was done by the coin. And this is really fun, because they say, "How can you possibly know this? It's all random." Well, let me just demonstrate it in this example.

Here are 200 random H's and T's, and here are human generated H's and T's. Now, of course, I know it just looks like an ocean of H's and T's to you, but what I look for is the following: I look for strings of long sequences where there all H's in a row or all T's in a row. Here we go. In the 200 randomly generated H's and T's you can see that I've colored sequences of H's here, and T's, that were at least four or five long, and look at this long sequence of H's. See that long thing? That was all by randomness alone. Look—six T's in a row here; five T's here; here are six T's in a row—a lot of streaks of many things in a row.

Now, a human being: How often will a human being write more than four strings of the same letter in a row when they're trying to be random? Well, they sort of resist that, because they don't think that's very random. They think you've got to sort of alternate—H-T-H-T— and so here in this human generated one you can see there are very few strings of H's and T's in a row.

Well, as a matter of fact, the probability when you flip a coin 200 times, the probability of having at least one string of 6 or longer of H's or T's is like 96%—very likely. And the probability of having at least one string of five is 99.9%—it's essentially certain. You'd be very unlikely to flip a coin that many times without getting these long strings, and in fact, if you actually simulate this on the computer, you'll see that that plays out, that you just almost always get these strings.

Well, we're looking then for counter-intuitive features that are presented by randomness, and this next one is one of the common misconceptions that a lot of people have about randomness, and that is that, suppose that you're flipping a coin again, and this time we're going to flip a coin, we'll flip a coin many times, and suppose that just randomly it happened to come out that 10 times in a row you got heads—10 times in a row you got heads. Don't you sort of feel in your heart that that coin is wanting to make a tails next time? Doesn't it feel like the next time it's more apt to be a tails? And the answer is, of course, that the coin doesn't know what it's just done. To the coin, every flip is a new flip, and it's just as likely to be a heads as a tails after it's done 10 heads in a row, as it was to get a heads than a tails if it had done none of them.

Now, let me show you what we did to demonstrate this. What we did is we simulated the following experiment, and that is, we took a coin, and over a million times, a million times, what we did is we flipped the coin 11 times. You follow me? In other words, we took a coin, and with a computer—computers are great, by the way; they don't care; a million times, they'll just go ahead and do it—so you just do it a million times, and we said, "What do you get?" And we wrote down this screen—H-T-T-T-H and so on—11 in a row. Then we did it 11 more times, 11 more, and we did that 11's 1,024,000 times. The reason we picked 1,024,000 is because every 1,024 times, that's the probability of getting 10 heads in a row. That'll happen, on average, one out of 1,024 times. So in other words if you do the experiment of flipping the coin 1,024,000 times, and each time you flip it 11 times, you expect that the first 10 will all be heads about 1,000 times. About 1,000 times.

Well, we did that experiment, the experiment of flipping 1,024,000 times, 11 times each one, and what did we get? Well, the number of times we got 10 heads in the first simulation was 1,008. You know, we expected 1,000: 1,008—extremely close. What happened to the 11th coin? Well, 521 times it turned out to be a head also. And 487 times it turned out to be a tail. You see, there's no memory. About half the time heads; half the time tails.

We did it again—983 times it came out heads; the first 10 times in a row were heads, and then the next time 473 heads, 510 tails. A third experiment, 1,031 times it came out heads 10 times in a row, and of those, 502 had the next coin be a heads, and 529 a tails. So you can see that the coin has no memory. After it's gotten 10 heads in a row, it's just as likely to be heads the next time as it was the first time you flipped that coin.

Okay. Now, I wanted to talk about another counter-intuitive aspect of probability, and it's really interesting to think about what is rare, and how we view rarity in probability. Suppose you got dealt the following hand: the two of spades, the nine of spades, the jack of clubs, the eight of spades, and the five of hearts. Well, you could look at that hand and say, "Wow, the two of spades, the nine of spades, the jack of clubs, the eight of spades, the five of hearts. I'm going to write home to my mother and to my friends and relatives and say, 'I got the most amazing hand today. It was the two of spades, the nine of spades, the jack of clubs, the eight of spades and the five of hearts. That's amazing. The chance of getting that particular hand—do you know what the probability of getting that hand is? One out of 2,598,960—that's the probability of getting that hand.'"

Now here's another hand; here's another hand. This is the ace, king, queen, jack, ten of spades. This is a royal flush in spades. What's the probability of getting this royal flush in spades? Exactly the same—1 out of 2,598,960—and yet you would write home to your mother about this hand for sure, and this one you wouldn't write to anybody. This is just an average hand, and yet in your whole life of playing cards, you know what? You will probably never get that hand again, because its probability is almost zero—1 out of 2,598,960. So this is one of the counter-intuitive concepts of probability: that rare events happen all the time, but you may not recognize them as of significance.

Okay. Now, let's look at some other rare events. Rare events absolutely happen by chance alone. The most common rare event that you see mentioned in the newspapers every day is the lottery. The probability of winning the Powerball lottery is about 1 out of 146,000,000. This is the big multi-state lottery in some states. One out of 146,000,000. That chance is so remote you'd think it would never happen; but it happens regularly. Why? Because a lot of people try. A lot of people buy random numbers and some of them then occasionally win. If you try something that's rare often enough, then it will actually come to pass.

This concept—that rare things will actually happen if you repeat them enough and you look for them enough—was encapsulated in an observation that was first made by the astronomer Sir Arthur Eddington in 1929, and he was describing some features of the second law of thermodynamics, and he wrote the following. He said:

> If I let my fingers wander idly over the keys of a typewriter it might happen that my screed made an intelligible sentence. If an army of monkeys were strumming on typewriters they might write all the books in the British Museum. The chance of their doing so is decidedly more favourable than the chance of the molecules returning to one half of the vessel. [Talking about the second law of thermodynamics.]

Well, so this brought up the question of, if you let monkeys randomly type on a typewriter, they will eventually write the entire script of *Hamlet*. They'll get it exactly right. So *Hamlet* happens. This is, this is—No. I wanted to talk about these monkeys typing on the typewriter. It is true. You see, the reason that it is certain to happen if you let those monkeys type long enough is the following: There's some chance they'll get the first letter right, and then having gotten the first letter right; there's some chance they'll get the second letter right; and then there's some chance of getting the third letter right. The probability of getting all those right is—well, if 1 out of 100 times they get the first one right, then of that 1 out of 100, 1 out of 100 times they'll get the second one right, and 1 out of 100 they'll get the third one right, and so on—and so there is some non-zero chance that they'll get it right every single time. And if they're going forever, well forever is a long, long time, and they will eventually get it right.

But I thought it was sort of amusing to see how much you could really expect to get. Suppose that you tried to type, "To be or not to be." That has 18 characters. And suppose you have a keyboard that has 100 keys. So there are 100 to the 18th—$100 \times 100 \times 100, \ldots$ 18 times—different 18 character patterns that you could write. That is, 1 with 36 zeros after it. That's the number of different patterns. So here's the question; suppose that you tried the following thing: That at the moment of Big Bang, you started trying to type "To be or not to be" by randomness. You type these 18 characters, you tried 18 characters, and then you'd say, "Is it right or not?" and then you'd try another 18 characters, and so on, and you did a billion of those per second—a billion trials per second—from the beginning of the Big Bang until now. What would be the probability that you would have typed just the phrase, "To be or not to be"? Answer: Your probability is about one in a billion that you would have even got to "To be or not to be" by this time, much less the complete works of Shakespeare. So I thought it was sort of amusing to see what the actual answer is.

However, you can find patterns in random writing, and in fact this happened, and a person made a lot of money by this. An enterprising author made a lot of money when he wrote *The Bible Code* a few years ago. You may remember *The Bible Code*, and what he did is they took the Bible, written in Hebrew, and they found patterns of words by skipping a certain number of letters and then finding that by skipping a certain number of letters, in that pattern of skips they would find words written out. One of them was like "Atomic holocaust Japan 1945." They found this by taking letters that were spaced out evenly in the Bible, and they said this is an example of showing how the Bible was showing the future.

Well, of course, the truth is that this is just a matter of probability. If you take all possible sequences of different lengths, you can by randomness alone find surprising things, and here just to demonstrate it, people found in debunking this analysis, they found patterns in *War and Peace* and so on. But since this is a mathematics class: They also found a very famous mathematician, namely, the Unabomber, Ted Kaczynski. You may remember him, he mailed bombs to people, and he wrote a manifesto, and so they used the strategy of *The Bible Code* to find things in the manifesto of Ted Kaczynski, and here's what they found.

They took the letters of the manifesto, putting them together, and they used the same methods that were used in *The Bible Code* analysis—which involved, by the way, they would string the letters together, but then they would just cut out little squares from it, so it gave them a lot more flexibility of finding patterns. And in this Unabomber writing, this part of it, look what you find—"Bomb Mail"—in this nice organized pattern, just exactly like they found in *The Bible Code* kind of analysis. This is another challenging part of probability, and that is that if you look for rare things but you have a lot of places to look, you'll tend to find them.

There are many challenges, I think, to randomness. One of them is that looking at things in retrospect, you look back at things that appear to be random, and yet you can explain them in retrospect. For example, stock movements; this is something we see all the time, but there are lots of people who can tell us exactly why the stock market moved in exactly the way it did from last week to this week, or from yesterday to today. There's no end to that; then you wonder, "Well, jeez, why don't they just tell me what's going to happen tomorrow, and then we could all be rich, you see?" They don't do that, because you can explain randomness; you can find out the actual things that happen to cause things that are better described randomly. Even rare events happen randomly, as we've seen before.

In fact, in the case of the stock market, psychics and others of that ilk can often be correct in their prediction of stock market predictions because if there are enough psychics, they'll guess all sorts of things, and randomly some of them will be correct.

Well, it's not always easy to detect what the difference is between something that was created by a random process versus something that is more determined by some process. One of the challenges is, how can we tell the difference? For example, if we had a sequence of H's and T's, as we did before, how could we tell whether or not that was created by a random process or by a non-random process?

Well, we already talked about the fact that if you flip a lot of coins you would expect to have a certain number of sequences that were a certain length, and in fact that kind of analysis, of saying, "Well, first of all, you expect about half heads and about half tails." The law of large numbers tells us that as we take more and more flips of the coin, we would expect that fraction of heads to tails to become closer and closer to even, and the fraction becomes closer to 50%.

Then you can be more refined and say, "Well, what number of HH's in a row would you expect? What fraction of those would you find in the whole pattern?" What fractions of TT's? What fractions of four H's in a row, and so on. Each one you can compute what the probability is, and then you can look at your list of H's and T's, see whether or not the fraction is somewhat what would be predicted by randomness, and then that would you give you evidence that the pattern is or is not generated by a random process.

There are some philosophical challenges here, and I'm going to leave you with one of the ones that I think is very difficult to think about; namely, the digits of the number π—π as you know, is the ratio of the diameter of the circle to its circumference. That is, the circumference over the diameter is exactly the number π, and the number π is 3.1415926 and so on, and it goes on forever. It has infinitely many digits if you got it exactly right, and it's been computed to billions and billions of digits.

There's nothing random about the digits of π. It's absolutely determined—every single digit—and yet if you look at the digits of π and you try to analyze them from the point of view of randomness, you find that it has a lot of the characteristics of randomness. Here, for example, is a chart that shows us the distribution in the first

10,000 digits of π, of how many times there is a 0, a 1, a 2, a 3, a 4, among the digits of π, and you can see that it's a very even kind of distribution, as you would expect from a random list of numbers—just randomly choosing digits from 0 to 9 and putting them in order. And likewise, if you look at other patterns, how many groupings of numbers that appear together, and so on, π satisfies a lot of those conditions.

Well, the challenge of looking at and asking what is random in the world. Is it random that a basketball player gets a certain number of free throws in a row, or is there some underlying cause? These kinds of questions are a real challenge to us. When we think about analyzing the world, which things we want to describe as random and which things we want to describe as deterministic—that is, cause and effect of something that doesn't have randomness in it—those kinds of issues present us with a real philosophical challenge.

In the next lecture we're going to be talking about the concept of expected value, which is way of taking probabilistic situations and giving a measurement to the value of different outcomes in order to make the decisions. I look forward to seeing you then.

Lecture Three
Expected Value—You Can Bet on It

Scope:

When we bet money in a gambling game, such as roulette, we know the probability of winning, and we know what our winnings will be if we win. We do not know, however, the specific outcome. If we repeated that exact bet millions of times, we would win a predictable fraction of the time; thus, the average win or loss per bet is a predictable expectation over the long haul. That is to say, while we do not have deterministic regularity, we have statistical regularity. This average win or loss is called the *expected value*. As we saw in the last lecture, the Law of Large Numbers tells us that as random trials are repeated more and more, the fraction of times that a particular outcome occurs will more accurately reflect the probability of that outcome, and thus, the actual average win or loss per bet will become close to the expected value. The concept of expected value allows us to assess the wisdom of various random enterprises that have payoffs or consequences. Betting on red in roulette, buying insurance, or buying a lottery ticket are all susceptible to expected-value analysis. As is common with probability topics, expected-value considerations lead us to some interestingly paradoxical situations. Expected value is our first attempt to understand what kind of regularity these probabilistic experiments have.

Outline

I. Many daily-life decisions involve randomness.

    A. Buying stock, having surgery, studying for a test, and buying insurance all involve making such decisions.

    B. How do we make these decisions?

        1. We consider hypotheticals and perform a sort of "cost-benefit analysis" for each possible outcome.

        2. One math strategy is to start with ordinary thinking and abstract it. As Albert Einstein said, "The whole of mathematics is nothing more than a refinement of everyday thinking."

3. We need to balance the likelihood of the various outcomes with the cost or benefit of each, which leads to the concept of *expected value*.

II. Let's use gambling, specifically roulette, to look further at this concept.

   A. There are 38 possible outcomes in American roulette. Betting $10 on a single number will pay $360 for a winning bet.

   B. The probability of winning is $\frac{1}{38}$; thus, if we place a bet 38,000 times on 13, we should win about 1,000 times (and lose 37,000 times).
   1. Therefore, we should win a total of $360,000.
   2. However, we would have paid out $380,000.
   3. Our loss is $20,000; per bet, the average loss is –$20,000 divided by 38,000 bets, or –$0.53. Hence, the expected value of the $10 roulette bet is –$0.53.
   4. *On average*, the bettor will lose 53 cents per bet.

   C. Expected value is an average.
   1. We have a collection of outcomes, and we have a probability for each outcome's occurring.
   2. Each outcome has a value associated with it. In this case, for 13, the value is $350, and for the other 37 numbers, it is –$10 (the money bet on the non-winning number).
   3. Let $O_1, O_2, O_3, \dots$ denote the possible outcomes.
   4. Let $P(O)$ denote the probability of an outcome and $V(O)$ denote the value of an outcome.
   5. Then, the expected value is: $P(O_1)V(O_1) + P(O_2)V(O_2) + P(O_3)V(O_4) + \dots$ and so on through however many possible outcomes you have.

   D. If you bet $10 on red, your chances of winning are $\frac{18}{38}$ and of losing are $\frac{20}{38}$.
   1. The payout of a $10 bet on red is $20, for a gain of $10.

2. Therefore, the expected value of the $10 bet is:

$$\frac{18}{38}(\$10) + \frac{20}{38}(-\$10) = -\$0.53$$

E. Casinos count on the Law of Large Numbers to ensure their profits, as the table of roulette simulations illustrates.

Repetition	Average Gain in 10,000 Bets	Average Gain in 1,000,000 Bets
1	−0.41	−0.50
2	−0.66	−0.54
3	−0.56	−0.52
4	−0.65	−0.52
5	−0.41	−0.51
6	−0.70	−0.55
7	−0.56	−0.52
8	−0.44	−0.52
9	−0.51	−0.53
10	−0.58	−0.54

1. When we made 10 repetitions of 10,000 bets on red, the average is very close to the predicted average loss of −$0.53.
2. When we made 10 repetitions of 1,000,000 bets on red, the average is even closer to the predicted loss of −$0.53.

III. Let us look at unexpected instances of expected value.

A. Suppose someone plays roulette 35 times, betting on one number each time. The expected value of each bet is −$0.53. And the expected total value of the 35 rounds = 35 $\left(\frac{1}{38}350 + \frac{37}{38}(-10) \right)$, or −$18.42.

1. Surprisingly, the probability that a bettor would be ahead after 35 rounds is $1 - \left(\frac{37}{38} \right)^{35}$, or 0.61.

2. However, the bettors who are ahead are only slightly ahead, and the people who are behind have lost $350.

3. Because the expected value gives weight, the expected value is negative.

B. Here is another example. Let's say you own a pub and you have a dart game with four rings.

   1. You wish to have the payoff be $4 for hitting the inner circle, $3 for the next largest ring, $2 for the next largest, and $1 for hitting the large outermost ring.

   2. You assume anyone who throws the dart has an equal chance to hit anywhere.

   3. You calculate the area of each ring and find that the largest has 44% of the area, 31% for the second largest, 19% for the third, and 6% of the area is in the small center.

   4. You can calculate the average expected payoff: $0.06 \times \$4 + 0.19 \times \$3 + 0.31 \times \$2 + 0.44 \times \$1 = \$1.87$

   5. If you decided to make the game completely fair, you would charge $1.87 per dart thrown, because a fair game is one where the expected value is 0.

IV. Let us consider another unexpected surprise in dealing with the expected value.

A. What is the expected number of rolls of a die until a five appears?

   1. If we roll the die 6,000 times, we expect 1,000 of those rolls to result in a five.

   2. The simulation results are very close to 1,000.

B. Now we ask what the average gap is between fives in that long list of 6,000 numbers.

   1. The answer is 6.

   2. We have 6,000 numbers, around 1,000 of which are fives.

C. But what if we take the long list of 6,000 numbers and randomly choose any point on that list and ask ourselves what the gap is between fives? What is the expected value of the length of the gap (the number of spaces between two consecutive fives)?

   1. The answer is 11, not 6.

2. The reason the answer comes out bigger than 6 is that we are more likely to choose long intervals than short intervals.

3. Likewise, if we cut a string to represent the various lengths on the list between fives and mix the pieces in an urn, we are more likely to choose a longer piece from the urn than a shorter one.

Readings:

Edward B. Burger and Michael Starbird, *The Heart of Mathematics: An invitation to effective thinking*, 2nd ed.

Sheldon Ross, *A First Course in Probability*.

Questions to Consider:

1. Suppose you play a game with a weighted coin that lands heads up $\frac{2}{3}$ of the time and tails up $\frac{1}{3}$ of the time. If you are paid \$6 if it lands heads and \$4 if it lands tails, what is the expected value of playing the game once?

2. Expected value does not mean that the expected value is what will happen. When lotteries have very high prizes, the expected value of buying a \$1 lottery ticket can be \$2 or more. Even under those circumstances, why is it not a good investment for you to mortgage your house and buy lottery tickets?

Lecture Three—Transcript
Expected Value—You Can Bet on It

Welcome back. Many life decisions that we make every day involve randomness and outcomes that we just really cannot predict. For example, if we buy a stock, we don't know whether the price will go up or go down. If we decide whether or not to get eye surgery that will correct our vision, well, several things could happen; one is that everything could work out great and we'd have perfect vision, or things could not go so well. There are consequences to different alternatives of the future, and we have to sort of weigh them.

You know, I teach in the university, and students constantly make the decision of whether they should study for a test or just rely on knowing those three things that will be asked on the test. You see this is a trade-off they have to make.

Or when we buy insurance; we buy insurance with the idea that something might happen, in which case the insurance will pay off, or maybe we're paying for insurance that's never used.

Well, how do we go about making these kinds of decisions? One strategy, a basic kind of strategy that we use in everyday life, is that we consider hypotheticals. We say, "Well, okay, it might happen that a hurricane comes and we have a disaster in our house, or our house burns down, and therefore it would be good to have insurance to pay for it." And if the house burns down, it would cost this amount of money, and it would be this much trouble, and therefore we say, "Well, it might be worth it to buy so much insurance." But it's sort of a sliding scale, and we have to give a sort of a cost-benefit analysis for each possible outcome, and we don't know which outcome's going to happen, and that's why there's a decision that involves something that's random. We don't know what's going to happen.

Well, one of the main strategies by which mathematics comes up with new ideas and develops things is to take ordinary thinking and abstract it, and generalize it, and sort of clarify what that thinking is really involved with. In fact, Albert Einstein has a great quote on this that I think is completely correct. He said, "The whole of mathematics is nothing more than a refinement of everyday thinking." And certainly in today's lecture that's what we're going to be doing. We're going to take everyday thinking and try to refine it to develop a concept.

What the concept is is that we're imagining that we have something that has several possible outcomes, and each outcome has a possibility to it, a probability to it, and then each outcome has a value associated with it. What we need to do is balance the likelihood of the various outcomes with the costs or the benefits of each of those outcomes. What we're really trying to pin down is this intuitive and natural thing that we frequently do, which is that we give more weight to things that are more likely. If something is very likely to happen, then we say, "Okay, I'd better consider that to be a much more serious possibility." Then if there is something that's extremely rare we say, "Boy, I sure wouldn't want that to happen, but it's so rare I can more or less ignore it."

Well, this concept of trying to pin down the idea of balancing the expectation—the likelihood of an outcome—with the cost or the benefit of that outcome leads to the concept of expected value, and that is the concept that we will define and discuss during this lecture.

Here is the idea: that we imagine having various outcomes, and as is often the case in the realm of probability, the best examples to illustrate a point are to start in the realm of gambling. In this case what we're going to do is use the gambling game of roulette to explain the concept of expected value, to illustrate this concept.

Here we go. A roulette table here consists of the following game. There's a roulette wheel that a ball rolls around and lands in a slot, and it lands in a slot that has a number, and the number is any one of the numbers 1, 2, 3 … up to 36, or there's also a 0 and 00. The probability of any one of the numbers from 1 to 36 plus 0 and 00, for a total of 38 possible outcomes, are equally likely when you roll the ball on a roulette wheel; or it's supposed to be equally likely.

What happens is the way that the payoffs are done in roulette is the following: That if you make a bet, you put a certain dollar bet on a particular number. So let's suppose this chip represents a $10 bet, and if you put the $10 bet on number 13, then the wheel rolls around, and if the number 13 comes up on the wheel, then the payoff is $360. In other words, you get a total of $360 back. Well, the probability of getting a 13, the 13 coming up, is 1 out of 38, because there are 38 equally likely numbers to come up, and the number 13 is just one of them. Your probability of winning is 1 out of 38, but when you do win, you win $360; that is, you take back $360, which is $350 more than you invested—if "invest" is the correct term.

So the concept of expected value is to consider what the average winnings or losings would be if you repeated the experiment many, many, many times. An expected value is an average winning or losing of the particular trial or experiment that you're conducting. In the case of betting $10 on the number 13, let's just see what it would mean to say: What is the average winning or losing that you would expect from undertaking the bet of betting $10 on number 13?

Well, here's what we'll do. Let's imagine that we conducted this experiment many, many times, and for convenience, let's imagine that we do it 38,000 times; and we reason we do it 38,000 times is because there are 38 equally likely numbers to come up, so that means that on average we would expect 1,000 of the bets to come up 13. If we did 38,000 times, we'd expect to win 1,000 times. Of course, it wouldn't be exactly 1,000 times, but on average it would 1,000 times; and we would expect to lose, by the way, 37,000 times of this 38,000 times that we bet on number 13.

In making those 38,000 bets, if we win 1,000 times the total payoff, the total amount of money that we take in is the $360 that we get whenever we win times the 1,000 times that we win, for a total of $360,000 that we take in. But we pay in order to have played 38,000 times—we pay $10 per time—so that's a total of $380,000 that we paid out. Well, the difference is the difference between $360,000 that we brought in and $380,000 that we took out, of course: We lost $20,000. That means that per bet the expected value of that bet is just a fraction of the $20,000 loss divided by the 38,000 bets that we made. In other words, it's the average. Over all of the bets that we made, it's the average winning or losing. In this case the average is – $0.53; in other words, on average, we expect to lose 53 cents for every $10 bet. So the expected value of making a $10 bet is –$0.53.

The first principle to note about expected value is that it's an average. In no case is it exactly the amount that you're going to win or lose. There's no way that you can bet $10 on number 13 and have them lose exactly 53 cents. That doesn't happen. But on average, if you did it many, many times, 53 cents on average would be the amount that you lose.

Now, by the way, another way to look at how this probability is computed is that it's equivalent to saying that 1 out of 38 times—that's the probability that you're going to win this bet when you put it on number 13; it's 1 out of 38—with that probability, you will win $350; that is to say, you will be ahead by $350. Whereas, 37 out of 38 times—that's the probability of some other number than 13 coming up—you will lose $10. And if we do this computation, we get the same answer: –$0.53.

That is the definition, then, of expected value. The following is the formal definition: If we have a collection of outcomes—in the case of the roulette wheel there were two outcomes that we considered; one was it comes up number 13, which we bet on; and the other outcome was that it comes up some other number. They don't have to be outcomes that have an equal likelihood of coming up. In the case of the roulette wheel, there was only a 1 out of 38 chance that it would come up 13; whereas there's a 37 out of 38 probability that it comes up some other number.

We list the potential outcomes of the experiment or trial, and for each one we have a probability of that outcome's occurring. For every one of those outcomes we also have a value associated with it. In other words, if that outcome happens, like coming up with a 13, there's a certain value associated with it. In this case it's +$350. You're $350 ahead. The value of the other possibility—some number other than 13—was that you come out behind by $10. So the computation of expected value is the sum of a product. It is the probability of an outcome times its value, plus the probability of the second outcome times its value, plus the probability of the third outcome times its value, and so on throughout however many outcomes there happen to be in the experiment you're undertaking.

Let's do a modified experiment, another roulette experiment, to just demonstrate this concept of expected value. We're going to do a couple of these so that you really understand the concept of expected value and how you compute it. Another kind of bet you can make on roulette is a bet on red, and this is what this red segment here illustrates. If you bet your $10 on red, what that means is that you're going to win if any of the red numbers that you see on the board comes up. Now, there are of the 36 numbers from 1 to 36, 18 of them are red and 18 of them are black. The two numbers 0 and 00 are not either red or black. So your probability of coming up with a red number is 18 out of 38 possible numbers, so you will win with a probability of 18 out of 38, and you will lose with a probability of 20 out of 38.

Now the payout for this $10 bet is $20; in other words, you gain $10. If you bet on red and you get a red, then you get paid $20; you've invested $10, so your net gain is $10. Well, we can compute, then, the expected value of this $10 bet on red; namely, it's the probability—18 out of 38—of getting a red, of a red coming up; times the amount that you gain, namely $10; plus the probability, 20 out of 38, that some other number that's not red, a non-red number, comes up. Twenty out of 38 are non-red, and in those cases you lose $10. If you just multiply these values out, you see that once again you have exactly the same expected value of that $10 bet: –$0.53.

In both cases the expected value of a $10 bet on red is exactly the same as the expected value of a $10 bet on an individual number. And by the way, what does that mean? Well, that means that if you actually repeated the bets millions of times, on average in each case you would lose $0.53 per bet, regardless of whether you bet on 13 or you bet on red. The expected value over the long term is exactly the same.

Remember that the Law of Large Numbers tells us that if we do a lot of trials of something that has a certain probability that we can compute, then the number of times that the outcome is what is being predicted divided by the number of trials is going to converge to the probability. In the case of roulette, this is exactly what it is that the casinos count on.

We actually did some simulations to illustrate this, so here's what we did: We simulated the idea of making a $10 bet on red in roulette, and we did it 10,000 times. We made 10,000 bets on red, and we calculated the average gain—which, of course was actually a loss— the average loss, and we did that whole simulation 10 times. So 10,000 times we bet on red, and we just saw what the average gain or loss was. Then we tried the same thing—we bet $10 on red 1,000,000 times; we repeated that 10 times. This is what we did. This is what computers are great at, of course. They're perfectly happy as a pig in a sty to just be sitting there and just simulating a million times of betting on red. And look what we see.

After betting 10,000 times on red, the average bets you see in this column, the average loss varied from I guess the low here was 41 cents and the high was a remarkable 70 cents, but you see that the average looks to be very close to the 53 cent average loss that we had computed from our expected value analysis. And look what happens when we did it a million times. We bet on red a million times by simulation. Look at those numbers. Look at how close they are to 53 cents average loss, after we did a million times, once again illustrating the concept of the law of large numbers, that when you repeat something many, many times, the computation of what the probability is and what the expected value is comes out to be what you actually observe.

Now why is this important for casinos? Well, casinos are very happy to do this kind of calculation because they are, in fact, experiencing this repeated process of making bets on numbers. Namely, all the people who come in and play roulette, they're all betting these numbers, and the casino is sitting there happily saying, "You know what? Some people are going to win; some people are going to lose; but on average over a lot of bets, and I know a lot of people are in there betting, on average I can be pretty confident that I'm going to earn" I, being the casino, "I'm going to earn 53 cents per $10 bet on average over the life of playing this game." And so this is a terrific thing for the casinos, and they can be very clear on their prediction of how much money they're going to make given how much money is invested in gambling. And so this is the way casinos operate.

Now I want to show you a couple of sort of surprising things about expected value, sort of a little bit unexpected, instances of unexpected expected value. So here's an example: Suppose that you have this roulette game, and instead of playing it just once, you play 35 times. In other words, you make a bet, you see what happens, and then you bet 10 more dollars and you see what happens, and you do it 35 times—you play 35 times.

Well, we know what the expected value of playing 35 times is. We know that on average we lose 53 cents per bet. There are 35 rounds. We can just do this computation here, and see that our expected loss is going to be $18.42. That's the expected loss over these 35 rounds of betting. On average we'll lose $18.42.

But here's what's surprising. What's surprising is that if you had a bunch of people making these 35 bets, in fact 61% of those people will be ahead after 35 rounds. Sixty-one percent, more than half of the people, will actually be ahead. In other words, suppose you had 1,000 people, each one did it 35 times, 61% of those people—over 600 people—would actually be ahead after 35 rounds.

Now, this seems to be a contradiction, because we just computed that the average loss was eighteen dollars and something. How could it be that 61% of the people, more than half, were actually ahead? Well, the answer is that the people who are ahead on average just won one time. Most of the people who are ahead, they're just slightly ahead, because, you see, they won just one time; but the people who are behind have actually lost $350.

Now, of course, some people won more than once—you know that could happen—but most people just win once. So the balance is that the losers are losing more money to compensate for the winners winning a little bit of money. By the way, in order to make this computation of the 61%—it's a simple computation, namely this: We just say, "What is the probability of losing every single time in making 35 bets?"

Well, the probability of losing any one time is 37 out of 38. That's the probability if you bet on a single number: Your probability of losing is 37 out of 38. And so if you then bet again, your probability of losing again is 37 out of 38 × 37 out of 38, because of the times you lost the first time, then you have to lose again the second time. And then the third time, 37 out of 38. So if you wanted to lose 35 times—you don't *want* to lose—but, to lose 35 times in a row, the probability is just 37/38 multiplied by itself 35 times. So that's the probability, this part right here, is the probability of losing all 35 times. So the probability of winning at least once is 1 minus that. And consequently, if you actually just put it in a calculator and do it, you see that $(1 - 37/38^{35})$ is equal to about 61%.

Okay, let's do one or two more examples to see the concept of expected value so we really pin down this concept. In this example suppose that you are a person, who owns a pub or a bar, and you're deciding to have a dart game in the pub, and you want to have the following payoff for the dart game. You have this target here, and $4 will be the payoff for getting in the middle, $3 for the next ring out, $2 for the next ring out, and $1 for the outer ring. Suppose that you assume, maybe by putting the dartboard far away, suppose you assume that anybody who throws the dart has an equal likelihood of landing anywhere on the board, and that if they miss the board, you let them try again; so they definitely hit somewhere on the board, but it's equally likely to hit anywhere.

Well, then, we can look at the probability of their hitting in the different rings by computing the percentage of area that are in the different rings, and here we've done so—44% of the area is in the outer ring, 31%, 19%, 6%—for the four different rings in percentage area. So to compute the probability of landing someplace, you'd say there's a 44% landing in the outer ring, a 31% in that next to the outer ring, and a 6% chance of landing in middle area. So it's just by area because we're assuming it's just random, so aiming doesn't have anything to do with this computation.

What is the expected payoff of this game if we pay out $4, $3, $2, and $1, respectively, for the different rings? Well, remember the definition of expected value is the probability times the payoff. Probability times the payoff plus the probability times the payoff. So in this case there was a 6% chance of getting a payoff of $4; there was a 19% chance—the yellow ring—19% chance of having a payoff of $3; a 31% chance, 0.31, of getting a payoff of $2; and a 44% chance of getting a payoff of $1. Just multiplying those together gives the expected value, which is $1.87, meaning that if you had thousands of people randomly throwing darts at this thing, and getting that payout, you would on average pay out $1.87 per customer—that is, per dart thrown.

If you were the bar owner and you wanted to decide what would make the game completely fair, so that you would expect to come out even—you wouldn't either pay out more money than you brought in or vice versa, having the people pay more than they've paid—you would charge $1.87 in order to have this be a completely fair game. In other words, the definition of a fair game is one where the expected value is exactly zero. In other words, a person pays an amount so that the payout on average is expected to be exactly the amount that they pay. So if you charge $1.87 for this dart game, on average you would all come out even; everybody would be happy.

Now, roulette is not a fair game, as we saw, because the expected value is a negative number; and of course, casinos would never have a fair game because then they wouldn't have the expectation of winning money.

I'd like to finish this lecture by looking at an unexpected example of expected value—the unexpected in expected value. Here's what we're going to do; we're simply going to take a die and roll it many, many times, and we're going to look for 5s. The way we're going to do this is to simulate it, of course, and here's the simulation where we roll the die 6,000 times, so of course we expect 1,000 of those to be 5s, roughly, and here on this left-hand column is the sequence of numbers simulating rolling the die 6,000 times, and at the top we record how many 5s we have. And we'll repeat the simulation. This time 997, and so on, number of 5s; always close to 1,000 numbers of 5s whenever we repeat the simulation.

Now what we'd like to do is to ask ourselves the question: What is the average gap between 5s in that long list of 6,000 numbers? Well, the answer is simple because since there are about 1,000 5s and they occur every sixth time on average, the average should be about 6 as the average distance between 5s, and indeed it is. We see in the simulations that we always get some number very close to 6.

Let me ask you then a slightly different question, and that is the following: Suppose that we take this long list of all of these numbers—6,000 repetitions of rolling a number—and just choose any point on that list between two numbers, and ask ourselves, "How long is the gap between 5s?" So, for example, if we pick this spot between these numbers, the gap is—well, let's see—the 5 is the second number to the right, the 5 is the seventh number to the left, so the gap has 8 units between the 5s where we find it. Now, here's my question for you: If you randomly choose a gap between this long list of 5s, what would you guess to be the average length of that gap? Well, you probably think that we've already answered the question, that the average length of the gap should be 6. Surprise! The expected value for the length of the gap should be, and is, 11. Not 6; it's 11.

Okay, let's see why. If you choose a number at random between—and what we've done here in this simulation is that this column shows us what the average of the gaps is if we randomly choose a point between values—so if we look here, for example, what this number represents is if we chose this gap between this 5 and this 3, then the total length of this gap is 5. Well, likewise if we choose the gap between the 3 and the 4, the length is 5 again, 5 again, 5 again. Well, why would the average of those things be 11 rather than 6?

Well, the way to see this is by thinking in a different way: namely, suppose that we imagine putting down all of these long list of 6,000 numbers, and taking a piece of string—this is red string—and we cut it to the lengths between the consecutive 5s. Then we put all of these in an urn here—it has to be an urn, of course, in order to be in probability class—you put it in an urn and then randomly choose a gap from the urn. Well, why are you more apt to choose a long string than a short string if you make such a choice?

The answer is because the long strings are more apt to be selected. For example, suppose you just had 2 strings in there; one had length 11" and one had length 1". Then if you reached into the urn and you chose any inch at equal probability, you would choose the 11" one 11 times to every time you chose the 1" one. So the average length of those strings would be closer to 11" than to 6", which is the average of the length of those two strings ($11 + 1 \div 2 = 6$). That's the reason that you're more apt to choose a long gap. If we look at this list of the gaps in our simulation, notice that whenever you have a gap—like this gap has length 9, there are nine 9s that appear, because all of those 9 gaps, each of them has a gap of 9, and so if we get to a long gap—here's one of 11; here's one of 16—and we're going to have sixteen 16s in a row. So we're more apt to pick the long gap than the short gap.

Another example of this would occur if you think about waiting for a bus. Suppose that you're in a city where the concept of the bus line number 5 is that it just appears 1/6 of the time, but just at random times; they don't go evenly spaced. On average, it's every 6 minutes, but it just appears on a random minute. Well then, if you go up to the bus stop and you ask somebody there, "When was the last time that a number 5 bus came?" and that's a certain number of minutes, and then you wait to see when the next number 5 bus comes, and you say, "Well, that's the gap in the length of time between buses." The expected value of that gap is 11, not 6, even though the bus comes every 6 minutes on average. So I think this is a really unexpected example of expected value.

In the next lecture we're going to be talking about random fluctuations by taking a random walk. I'll see you then.

Lecture Four
Random Thoughts on Random Walks

Scope:

Suppose you want to go for a walk, but you feel in a particularly indecisive mood. You decide to walk along a straight north-south road while letting fate decide your direction at each block. You take out a coin and flip it. If it is heads, you walk one block north; if tails, one block south. At each block, you make that random choice. The path you take is called a *random walk*. Many intriguing questions arise in this indeterminate perambulation: Will you ever return home? Will you ever venture 100 blocks away? The analysis of random walks helps us to analyze real-life situations, such as counting ballots during an election, and it explains the sad fate of persistent bettors known as the *gambler's ruin*.

Outline

I. This lecture addresses the phenomenon of random fluctuations.

    A. Examples of random fluctuations include the stock market, ballots in an election, coin flipping, genetic drift, and Brownian motion.

    B. The simplest example is the random walk.

        1. As we leave home (position 0), if we flip a coin and get heads, we go one block north (position 1); if we flip tails, we go one block south (position −1).

        2. When we have walked one block, we then flip the coin again and go another block north or south, depending on the result, and so forth. We can see this walk recorded on a graph.

    C. This random walk generates many intriguing questions, such as: How far away do you get? The answer is probabilistic because it depends on flips of a coin.

    D. Another question is: When we take a random walk, what is the probability that, from position 1, we will return to where we started?

1. To answer that question, we can compute as follows: $P = \frac{1}{2} + \frac{1}{2}Q$, in which P is the probability that starting at 1, the walk eventually gets back to 0, and Q is the probability that starting at 2, the random walk eventually gets to 0.
2. We can ask what the probability is that starting at position 2, we will return to where we started. To get from 2 to 0, the walk must first get to 1 (probability P), then eventually to 0 (probability P). Thus, $Q = P^2$, and we arrive at the equation: $P = \frac{1}{2} + \frac{1}{2}Q = \frac{1}{2} + \frac{1}{2}P^2$
3. Working out the equation leads to the result $P = 1$; the probability is 100% that we will indeed return to 0.
4. Although some random walks never return to where you began, the fraction of walks that have not returned becomes closer and closer to 0 as you take longer walks. Thus, the probability of never returning during an infinitely long walk is 0.

E. Another question is: What is the probability that we will eventually get 100 blocks away from where we started? The surprising answer is again $P = 1$.

F. The *gambler's ruin* is a variation of a random walk.
1. A gambler starts with $2,000. Each bet is $200, with even odds.
2. Let's say the game involves flipping a coin, with heads meaning the gambler wins and tails meaning the gambler loses.
3. As we have seen in the random walk, the probability = 1 that you will eventually get back to 0. This means that the gambler will eventually lose everything, even in a fair casino.

II. Bertrand's *ballot theorem* deals with an election between two candidates in which the winner, A, receives a votes, and the loser, B, receives b votes, where a (52) is greater than b (47).

A. Suppose the votes are tallied by drawing them out of the ballot box one by one, adding 1 to the proper person's score. What is the probability that the eventual winner will always be ahead, from the very first vote counted?

B. This problem can be rephrased as a graphical problem.

    1. Consider the graph whose horizontal axis is time (or ballot number) and whose vertical axis is the amount by which the eventual winner is ahead.

    2. The answer to the question of what the probability is that the eventual winner will always be ahead, from the very first vote counted, turns out to be $\dfrac{a-b}{a+b}$.

III. This discussion brings up the question of potential ties.

    A. Suppose you wish to hire a tennis pro. Two candidates have played one match against each other each day for the past year, keeping a running tally of how many matches each has won. The tally shows that one player was ahead for the entire last nine months, so that player seems to be better.

    B. By comparing this situation with randomness, we can test the strength of that conclusion. Knowing what to expect from randomness informs our interpretation of the results.

    C. Let us consider the case of randomness in which, for 366 days in a row, two people flip a coin to win or lose, and let's see where we might expect the last tie to occur.

        1. We find, in fact, a surprisingly high probability of one person being ahead for most of the year.

        2. In fact there is a $\dfrac{1}{2}$ probability of one person being ahead for the entire last half of the year and a $\dfrac{1}{3}$ probability of one person being ahead for the entire last nine months of the year—by luck alone.

IV. If our case were expanded to north-south-east-west, then we would have a two-dimensional random walk.

    A. Such a random walk has some interesting properties.

        1. We can ask again: What is the chance of returning to the origin? As with a one-dimensional random walk, the probability of returning is 1.

        2. However, the rate at which we return is not so quick. In the 30 simulations we ran, it took anywhere from just 4 steps up to more than 100,000,000 steps before we returned to the origin.

B. Peculiarly enough, in the case of a three-dimensional random walk, which allows up or down as an additional choice, we have only a 35% chance of returning to the origin.

Readings:

John Haigh, *Taking Chances: Winning with Probability.*

Questions to Consider:

1. Suppose as you finish grocery shopping there are two checkout counters open, and both seem to have an equally long line. You pick a line. Does it seem that more often than not you pick the slow line? How does the fact that ties are less frequent than our intuition would predict relate to this situation?

2. Suppose two people play a game where one flips a coin and the other guesses how it will land. If the person guesses correctly, the guesser gets $1 from the flipper; if the guesser is wrong, the flipper gets $1 from the guesser. Suppose the flipper starts out with $10 more than the guesser. What is the probability that at some time in the future, if they play forever, they will have equal amounts of money?

Lecture Four—Transcript
Random Thoughts on Random Walks

Welcome back. Life, of course, is the source of most mathematical ideas. We look at things that happen in the world, and then we try to abstract from those some principles that become the mathematics that we're trying to develop. This is certainly true in the case of talking about probability and randomness.

But in today's lecture we're going to be looking at some phenomena that we see in everyday life that involve random fluctuations. For example, think about the stock market. We look at how a price of a stock varies over time, and if we look particularly at short periods of time, minute by minute, we see that the stock price varies in sort of a random up-and-down kind of a way.

Consider an election evening when we're counting ballots from a ballot box—and perhaps instead of thinking of a national election where there are millions of ballots, suppose we just have a small ballot box where we literally are taking out one piece of paper at a time—and we count one ballot at a time. Well then, one candidate seems to be going ahead, and then the other candidate gets a few votes, and it goes down and up, and that's sort of a random kind of a process.

Or we just flip a coin. We take a coin and we flip it. We get a heads, then a tails, then a heads; and sometimes the heads get ahead, sometimes the tails get ahead, and that's a sort of a random variation between heads and tails.

In sciences, there are lots of these examples. For example, in biology—we'll be talking in two lectures from now about genetic drift, how the aspects of the traits of the population can vary just randomly from time to time.

Another example is Brownian motion. You may have heard about this, where a grain of pollen is sitting on top of the surface of water; it seems to move in a sort of a randomly fluctuating way.

All of these are examples that involve a process of just taking random kind of motion, and so what we're going to do, as mathematics is want to do, is to abstract all of those ideas into some simple, essential ingredient and then explore that. What are we going to think is the simplest model for this kind of random fluctuation? Well, we're going to call it—the name that it goes by—is a *random walk*. What we envision is the following: We decide to go for a walk, but we're in a particularly indecisive mood. Instead of having a particular destination, what we do is we just go outside our door, and we say, "I'm going to flip a coin, and if the coin turns out heads, then I'm going to go one block north. And then when I get there, I will flip the coin again, and if it's heads, I'll go another block north, but if it's tails, I'll go a block south."

And so in this fashion, every block I flip a coin and randomly decide to go one block north, or another block north, if it's heads; or one block south if it's tails. And in that way we have something called a random walk. Now, I'll show you a graphic to illustrate this sort of random perambulation. Here we start at the point 0 in our random walk, and we just take a step because we flipped the coin and it came out heads, we went one block to the north, and then here we flipped heads again, so it went another block north. We flipped heads again so it went one block north again; heads again, one more block north; another block north. Oh, there, we flipped finally a tails, and we went one block south; another block south because of another tails that we flipped, and so on. And this continues on and on, and we just zoom back and forth, back and forth in a sort of a jittery kind of a way.

Now, it's hard to see this description of a random walk, and consequently we're instead going to look at it with a different kind of a graphic that is easier to see, and the way that we'll see this new graphic is that we're going to expand the time out to the right, so that as we take our random walk, the steps are going up and down just as they were before, but we'll record them by also saying what time— that is, what number flip that we're taking—in order to be able to see better where we are at different times. So here we are; it's the same random walk, but we can spread it out now over time.

Here we'll speed it up, and you can see the random walk that we're taking. This is after 100 steps we get right to here; this is recording a 100-step random walk. I hope that you follow me that at each one of the 100 flips of the coin, we're recording where we are relative to our original position. At this point here we've flipped, for example, in this case six more heads than tails; at that point, after we've flipped about 30 heads or tails. When we're here at the axis that means that we've flipped exactly the same number of heads and tails at that point, because that's what it means to be back to the original position where we started.

This is the way that we're going to describe our random walk. Now, after we've thought about this random walk, it's a rather abstract kind of an idea, and so the first question that we need to ask ourselves is: Well, what kind of question should we be asking about this kind of random behavior? What kind of issue comes up? Well, one kind of issue is the question of where are we going to get to if we take this kind of a random pattern. Do we expect that we're just going to hover around the origin, that is the place that we started, and we're just going to sort of go back and forth and stay in a sort of bounded area, or will we sometimes go way off into the distance?

Of course the answers are all going to be probabilistic, since what we're doing is taking a phenomenon that involves chance; namely, just flipping a coin. We know that there's some chance we'll just go back and forth, back and forth, back and forth forever, but there's another chance that we'll go further out, and we have to ask ourselves some questions about what are the chances that we go a great distance, or that we hover around zero. One basic question that we can ask is: Well, what is the probability that at some point in the future we return back to where we started?

In the random walk that we saw demonstrated before, we saw that we did return, in fact several times, in that 100 steps. But what is the probability that we return? Will most of the time we just sort of head off and never get back to where we started, or will most of the time we get back to the beginning? That will be our first question that we undertake. What we're going to undertake is a demonstration of what the probability is that when we take a random walk we will, at some time in the future, return back to where we started.

Let's think about this. Well, first of all, when we flip our coin, the first step is going to take us to a distance 1 away from where we started. If we think of our blocks as being numbered with 0 as where we start, and then 1 as in the positive north direction and −1 is the negative south direction. Then after one step we're going to be at position 1 or position −1. So if we're trying to compute the probability of returning to where we began, after that first step, it's the same question as: What's the probability, starting at position 1, that we get back to position 0? Or −1 to get back to 0? But since they're symmetric, it's the same probability.

Let's assume that we start at position 1 and we're asking: What's the probability of getting back to position 0? Well, we can think about this in the following way: First of all, there's a 50% chance that we immediately return right back to 0, because if we shake a tails, we go south, and therefore we go from 1 back to 0. So the probability that starting at 1 that we get to 0, that probability P is equal to ½—which is the probability that we'd shake a tails—plus ½, which is the probability that we shake a heads, times Q, where Q is another probability. It's the probability that starting at 2, a random walk will eventually get to 0.

You see, we made negative progress; that is, we went to 2 when we were trying to get to 0. So we went to 2 and now we're asking: What's the probability of starting at 2 getting to 0? Well, here's then what we can think about. In order to get from 2 to 0, we first, of course, have to get back to 1. What's the probability that we'll get back to 1, starting at 2? Well, that's the same question that we were asking to begin with: What's the probability of going from a position to one block lower down? So that probability is P. That's the unknown probability P which we're striving to understand.

In order to get down from 2 to 0, we first have to get to 1, which we do with probability P, but then we know that the probability of getting from 1 back to the 0 is once again probability P. So that means that Q entails doing something that has probability P, getting to 1, and then again doing something that has probability P; namely, going from 1 to 0. So $Q = P^2$. Therefore, $P = ½ + ½ Q$, which is equal to $½ + ½ P^2$. Now just a little tiny bit of arithmetic, multiplying through by 2, gathering terms, and remembering how to factor this little equation, we see that in fact $P = 1$. What this says is that with probability 1, when we take our random walk, we will eventually return back to 0.

Now, this might bring up in your mind a question; namely, the question that you can say, "Well, wait a minute. Suppose that every time I flipped heads, every single time, and I just kept walking further and further away. I never return to 0. I thought you told me," you're asking yourself, "that if something has probability 1, then that means it'll happen every single time, and yet here I've demonstrated an example where it doesn't happen that you return back to the origin." The wrinkle is that here we're talking about a probability that involves infinitely many possibilities; that is to say, there are infinitely many random walks you can take if you just continue to flip forever, and although there are some that do never return to 0, the fraction of those that never return to 0 is smaller than any positive number, and therefore we say that the probability is 1 that we return to 0, even though it is possible that we won't return to 0. This is the wrinkle about infinity that involves the probability being 1 even though it is possible that you will never return; but in practice you will always return to 0.

The next question that we can ask is: Well, what's the probability that in flipping a coin we'll go a certain distance away, such as, will we ever get as far as 100 blocks away from where we started by just randomly flipping a coin? Well, the answer to that is something that we have actually thought about before, because now we realize with probability 1 we will return back to where we started, but we know that there is some non-zero chance that we'll flip 100 heads in a row.

There's always a non-zero chance that we'll flip 100 heads in a row. Remember the monkeys that we met last time? The monkeys that said that even very rare probabilities will happen if we go on forever, and there is a non-zero probability ($1/2^{100}$) that we will flip 100 heads in a row. Well, if with probability 1 we return to 0, then that means that with probability 1 we'll return to 0 again, and return to 0 again, and again, and again. In fact, infinitely many times we'll return to 0, and so we have infinitely many opportunities to flip our 100 heads in a row, and with probability 1, one of those times that'll happen.

That means that we're going to get to any distance. And by the way, of course, in fact it's possible to get to 100, much more likely to get to 100, not by starting at 0 and flipping 100 heads in a row. In fact, we would just sort of tend to oscillate and then get up to 100 by a sort of jiggly way. But the proof that it's definitely going to happen is by observing that you're going to be back at 0 infinitely many times, so you have infinitely many chances to do a rare thing, like flipping 100 heads in a row.

What we've seen then so far is that you will definitely, by just this random process, get to any distance away from the origin. Now an implication for this is something called the gambler's ruin. The gambler's ruin is the following scenario: Suppose that you're a gambler and you go to a casino and you start rather well with $2,000, and you're betting at this casino and each bet, let's say, is a $200 bet. And let's suppose that this casino, which doesn't exist in the real world, just gives completely fair odds—it's a fair game. And, in fact, let's suppose that the game just consists of flipping a coin, and if the coin comes up heads, then you win 200 more dollars—you get $200 additional money. You get your $200 back and you gain $200. And if it's tails, you lose $200.

You start playing this game, just flipping coins. Now this is a fair game, and so on average you don't win or lose, but you're taking a random walk in the amount of money that you have. Now, what have we just proved? Well, we just proved that with probability 1 you will get back to where you started, and with probability 1 you will actually lose $2,000 at some point. So with probability 1 you're going to go broke, and that is called the gambler's ruin—and notice, by the way, this is a gambler's ruin in a fair casino, which doesn't even exist. You're going to be ruined a lot faster in a casino where the casino is getting a certain percentage of take on the money; in other words, where the expected value is negative, which is the case in real casinos in such things as roulette.

But what we want to do is to simulate this a few times. Suppose that we start with a gambler, who starts with $2,000, and we just simulated making these fair bets—$200 a bet—and seeing how many bets it takes before our hapless gambler goes broke. In this case, the gambler went broke after 187 steps. In this next simulation, however, the gambler did not go broke; and after, that is, even after 1,000 steps the gambler did not go broke, in fact, now this gambler has, oh it looks like $6,000 or $7,000; so this was good. Now he'll go broke eventually, but he hasn't gone broke yet. This one went broke after 65 steps, after 47 steps, after 57 bets, after—oh, this person didn't go broke, and look at this. This gambler is making more than $10,000 after 1,000 steps; so this is great. He'll go broke later. Okay. This person will go broke after 159 steps, and so on, 151 steps. This one took 703 steps; that is, 703 bets before that gambler went broke. And so this is the sad story of the gambler's ruin.

Now, there's another example that we're going to talk about now, which has to do with counting ballots in a ballot box. This is what I described before. Suppose you have a ballot box and in the ballot box you have the ballots, which are voting for either candidate A or candidate B. And let's assume that candidate A actually wins 52–47. So that's the eventual outcome. Our question is: How likely is it that during the counting of the ballots one at a time, what's the probability that the eventual winner always stays ahead as you're tallying the votes one after another?

Let me make sure that you understand what we're doing. I'll show you an example of what could happen. Here's an example of where you start with counting the ballots. There are no ballots counted at the beginning, and then after one ballot is counted, that ballot went to candidate A, the eventual winner. The next ballot went to candidate A, the eventual winner; went up for a while, then candidate B got some votes. And so what we're recording here is the margin of A – B—A's votes minus B's votes—as you are counting the ballots. You can see that if it's above the 0 line, that means that during the intermediate counting of the ballots—for example, 20 ballots have been counted here; it says that there were 10 more ballots for the eventual winner, candidate A, than there were for candidate B. In other words, votes for candidate B are counted as a -1 and votes for candidate A are counted positively, and we're taking the accumulated total as we're counting the ballots.

This is an example of a graph that indicates the intermediate counts of a ballot in which the candidate A did in fact stay ahead the entire time. Now, here's one in which candidate B got some votes, appeared to be ahead at the beginning of the count, and then later, of course, A always wins, because we're assuming that A got 52 votes and B got 47 votes. So these are all examples. Here's an example: A started out ahead, but then B got a lot of votes, and so B did at some point appear to be ahead, but then finally, of course, A wins.

The question is: What's the probability of counting the votes and having A always be ahead during the entire voting? Well, it turns out that there's a very neat closed form answer to this question, and that is that if the candidate A gets a votes and candidate B gets b votes; and A is the winner, so a is bigger than b; then the probability of counting the votes—that is to say, among all possible ways of picking out those ballots, the probability that A will always be ahead throughout the counting is ($a - b$ over $a + b$). ($a + b$) is the total number of votes cast, and a is the number of votes that A, the winner gets, minus b, the number of votes the loser gets.

And you can see why at least this makes some sense; because suppose that A just won by a little tiny bit, and so a and b, maybe they just differ by 1 vote. A just won by 1 vote. Then you can see that $(a - b)$ would be a very small number, like 1. It would be 1 over the number of ballots cast; and that makes sense, because if the voting is very close, it would be rather rare for the winning candidate to always be ahead, because the other candidate got almost as many votes. Whereas, suppose it were a slaughter, and A got almost all the votes, and maybe B got 1 vote, then this number, $(a - b)$, would be just very close to $(a + b)$, and consequently we have a fraction that's very close to 1, so the probability would be almost 1, that is, almost certain, that the intermediate counts would always favor the eventual winner. This is called the Bertrand ballot theorem.

Well, one of the reasons that I like to think about the Bertrand ballot theorem is that it brings up this question of potential ties, and let me illustrate it in the following way. Suppose that you are a person who's hiring two people for a job, and the two people have had some sort of a measurement of who is better day-by-day for many days, in fact, for every day for an entire year. And you have the record. They did something. Maybe they were tennis players, and they played tennis against each other; or they were taking some sort of an exam and one was better than the other; and every day they competed, and you had in front of you a record of who was ahead in their cumulative wins over the other one for most of the time.

Suppose that in looking at this record you say, "Oh, look, this candidate A was ahead for the entire last half of the year." In other words, they had won more times than the other candidate for the entire last half of the year. Boy, that certainly seems to be a good recommendation for that candidate. Or maybe this candidate was ahead for the entire last nine months of the job. Always the cumulative record of the wins of the one candidate over the other candidate, that the one candidate had won more times than the other, for the entire last nine months. It seems like it's very strong evidence that that person is actually better.

But here's what we're going to do in order to test the value of that impression, and that is to compare it against randomness. You see, if we know what to expect by just a random kind of a contest, like flipping a coin every day, we can then compare the effect of what to expect from randomness about ties and things, and one being ahead than the other, compared to what actually we saw in the example of the people, and it might give us an impression about how to interpret the value of the evidence that one was ahead. And I found this surprising, and I hope that you find it surprising too.

Here's what we're going to do: We're going to look at some simulations of where the contest between these two candidates consists of flipping a coin; that is to say, it is completely random at every time whether one person wins or the other person wins. And we're going to investigate the question of doing this for 366 days; how many times of doing it 366 days, where do we expect the last tie to occur? In other words, is it common for the last tie—do they oscillate back and forth, and almost always the last tie is right near the end of the year? Or how often does it happen that just by luck alone the last tie is near the beginning of the year?

We just did a bunch of simulations, and here they are. Here's the first simulation. Now, what this means by simulation is that we just simulated flipping a coin 366 times, and keeping the running total of heads minus tails. In other words, whenever we get a cumulative answer of 0, that's a case for a tie. In this case the last tie occurred here at about number, oh, 290-something. This vertical bar indicates the location of the last tie, the last time the heads and tails were exactly equal, and then it continues on here—all of these are more tails than heads for the remaining part of the year. And then we record this, we're going to do the simulation 1,000 times, and we just make a mark down here to indicate where the last tie was in this first walk.

In the second walk, the last tie occurred much earlier in the year, at about the 130th day. Then over that the heads person was ahead the whole time; then we do the simulation, here it was somewhere in the middle of the year. This time the last tie was essentially right at the end. Here the last tie was in the middle, here the last tie was right at the beginning, and we're accumulating where the last ties were over each of our 1,000 simulations.

Now we're just going by 10s, and I'm going to run it rather fast so you can see growing up this histogram on the bottom is telling us after 1,000 simulations where the last tie occurred; and look at how surprising it is. First of all, the last ties often occurred right at the beginning of the year. That's in fact the most common individual place was that it just deviated right from the beginning and never tied again. Then it actually turns out that it's a symmetrical curve; in other words, that the probabilities are that the last tie occurring at the first day is exactly equal to the probability that they occur at the very last day, and that it's symmetrical, and that it has sort of a U-shape to it.

What that means is that on average half of the time that you just flip a coin, where there's no bias at all, half of the time the last tie will occur in the first half of the year—and in fact a third of the time the last tie will occur in the first three months of the year. To me the impact of this is that if you have, for example, two children who are being evaluated against each other by some method, and you see one child being ahead in this measurement for many, many times over the year, just by probability alone a third of the chance they might be exactly equal at the beginning. I think this has some impact to it in the way that we maybe measure the different qualities of people.

I want to conclude this lecture by talking about a case where we're viewing fluctuations not just in one dimension but in two dimensions. That is to say, we're going to take a random walk, but starting at the origin we'll by randomness choose to go either north, south, east, or west, randomly. I can simulate this random walk by this arrow; it's taking this random process at each place on the grid. We go right, left, up, or down, and we just continue in this way, and this draws out this rather interesting random perambulation around the plane, and there it goes.

Now, this is an interesting thing. First of all, it's interesting to just watch this thing, but it turns out that this kind of random walk in two dimensions has some very interesting properties to it. One question that we can ask about a two-dimensional random walk is exactly the question we asked for a one-dimensional random walk; namely, what's the chance of returning to the origin? Well, it turns out that the probability of returning is 1, but the rate at which we return is not always so quick.

Here are some examples of random walks—1,000 steps of a random walk. First of all, they draw these interesting figures. A thousand step random walk. Sometimes they hover around the origin, but look, sometimes they just seem to go off—this one just seems to go off—and often they don't come back to the origin any time soon. So these are interesting random walks. These are 50 random walks. Some of them just go off to the side.

Now, what I'd like to ask is the question: Although the probability of coming back to the origin is 1, how soon do you do so? Well, we did simulations of these two-dimensional random walks, and asked: What are the number of steps before we return? And in these 30 simulations, which we took up to 100,000,000 steps; notice that some of these, like this one, required 8,000,000 steps before it returned to the origin. This one required 36,000,000 steps before it returned to the origin. And these four over here, we stopped at 100,000,000 steps—it still had not returned to the origin.

In the case of a three-dimensional random walk it becomes even more peculiar; namely, that in a three-dimensional random walk, the probability of returning to where you started is not 1. In fact, it's only 35%.

Well, what we're going to do is, in the next lecture, we're going to see some applications of this concept of random walks, and other applications of probability in the physical world. I'll see you then.

Lecture Five
Probability Phenomena of Physics

Scope:

Quantum mechanics describes the location of a subatomic particle of physics as a probability distribution. Our intuition would prefer elementary particles to be more like tiny round balls, each of which is at some specific place at each moment. But quantum mechanics suggests that an electron has some chance of being at any location in the universe at any moment. Interestingly enough, Einstein philosophically opposed the probabilistic nature of quantum mechanics. Weather predictions give us probabilistic descriptions of our world that have more obvious consequences. We might read, "There is a 30% chance of rain tomorrow." Then our question becomes: What exactly does that mean?

Outline

I. The probabilistic analysis of random behavior lies at the heart of physical phenomena, from quantum mechanics to the weather.

    A. One of the most basic features of understanding the world is that physical matter is made up of atoms and molecules.

        1. At the turn of the 20th century, the scientific community was not so clear that atoms and molecules actually existed.

        2. It turned out that strong evidence for their existence was an application of probability, and one of the major players in that analysis was Albert Einstein.

    B. It was Einstein's theoretical work on Brownian motion that allowed experimentalists to do actual measurements that helped confirm the reality of atoms and molecules.

        1. Brownian motion was discovered in the early 1800s by botanist Robert Brown, who made microscopic observations of grains of pollen on the surface of water and noticed that these grains appeared to constantly and randomly move in a jittery way on the surface of the water.

        2. In his 1905 paper, Einstein hypothesized that Brownian motion was caused by actual atoms and molecules hitting the grains of pollen, impelling them to take a "random walk" on the surface of the liquid.

3. Einstein wrote down a formula that predicted what distance a piece of pollen would move on average per unit time.

4. Experiments accorded with Einstein's predictions and, thus, were strong evidence for the actual existence of atoms.

C. In a sense, Einstein's work encouraged the mode of reasoning that led to the inherently probabilistic nature of quantum mechanics.

1. In quantum mechanics, the most fundamental objects that make up matter are to be viewed, not as being in one location at one time, but instead, as having a probability of being anywhere.

2. Einstein never accepted the probabilistic nature of quantum mechanics.

D. Probability plays a central role in physical theories, from quantum mechanics up the ladder of different sizes of interacting matter to chemistry and into macroscopic matters, such as the weather, which is where we now turn our attention.

II. Suppose you read that there is a 30% chance of rain in your region tomorrow. What should that statement mean?

A. First, we need to dispose of the issue of *threshold*, that is, how much rain is rain. The answer is 0.01 of an inch.

B. Second, what does it mean that there is a 30% chance of rain *at one spot*? It means that on about 30 out of 100 days in which the weather circumstances are like they are today, you would expect at least 0.01 inch of rain *in that spot*.

III. The problem arises when we hear we have a 30% chance of rain *in a whole region*.

A. Because there are different points in the region, we must deal with these variations.

1. The simplest case is if the region is very small and very homogeneous in its character. In that case, the conditions

are indeed the same throughout the region for the 30% probability of rain.

2. In another case, though, you might have 50 acres out of a 100-acre region where the probability of rainfall is 40%, and in the other 50 acres, it is 20%. The expected value of the probability of rain for the whole region is 30%.

3. In another case, though, you might have 30 acres out of a 100-acre region where the probability of rainfall is 100%, and in the other 70 acres, it is 0%. Again, the expected value of the probability of rain for the whole region is 30%.

4. In another case, you might have 25 acres out of a 100-acre region where the probability of rainfall is 40%, 50 acres where the probability of rainfall is 30%, and in the other 25 acres, it is 20%. Again, the expected value of the probability of rain for the whole region is 30%.

B. The consequence to the above conclusions is that on average, the amount of the area that will get rain is 30%.

1. In other words, suppose for each day of 10 days, we knew how many square inches were rained on; we would add all these square inches, then divide by 10 to get 30%.

2. For example, let's use a familiar example where 30 acres out of a 100-acre region always get rain, and in the other 70 acres, it never rains. Again, on average, 30% of the region gets rain.

3. Suppose now that every point in the region has a 30% chance of rain. We can look at each tiny square inch and record rainfall for 10 days there. We would expect to have rain on 3 of those 10 days.

4. Expanding our region to 10 square inches over 10 days, we see that every square inch would expect to be rained on for 3 of the days. Thus, the number of square inches rained on averaged over the 10 days is 3 square inches, which is 30% of the total area.

5. We would get the same result if we looked at a situation where half the region gets 40% chance of rain and the other half, 20%.

IV. The definition of probability of precipitation is tricky.

A. The official definition from the National Weather Service is, at best, misleading: "Technically, the probability of precipitation (PoP) is defined as the likelihood of occurrence (expressed as a percent) of a measurable amount (.01 inch or more) of liquid precipitation (or the water equivalent of frozen precipitation) during a specified period of time *at any given point* in the forecast area. Forecasts are normally issued for 12-hour time periods."

B. The definition should be written: "Technically, the probability of precipitation (PoP) is defined as the likelihood of occurrence (expressed as a percent) of a measurable amount (.01 inch or more) of liquid precipitation (or the water equivalent of frozen precipitation) during a specified period of time *at a random point* in the forecast area. Forecasts are normally issued for 12-hour time periods."

C. A multiple-choice question given to the public to determine if they understand the phrase "The chance of rain is 30%" proves that most Americans do not understand the definition of probability of precipitation.

Readings:

Ian Stewart, *Does God Play Dice? The New Mathematics of Chaos.*

Questions to Consider:

1. A consequence of quantum physics is that there is a non-zero probability that the moon will spontaneously fall on our heads tomorrow. Why shouldn't we be worried about that possibility?

2. Currently, weather prediction is viewed as a probabilistic enterprise. Do you think that with better knowledge of weather patterns, the randomness will be removed and weather prediction will become deterministic? The theory of mathematical chaos suggests not.

Lecture Five—Transcript
Probability Phenomena of Physics

Welcome back. Today's lecture concerns the role of probability in descriptions of the physical world.

The probabilistic analysis of random behavior lies at the very heart of how we understand physical phenomena; from everything from quantum mechanics to the weather. Physics and chemistry describe the nature of matter and how matter interacts with itself; and one of the most basic features of understanding the world is that physical matter is made up of atoms and molecules. But at the turn of the 20th century, the scientific community was not so clear that atoms and molecules actually existed. Some scientists thought that atoms and molecules were to be viewed merely as metaphorical concepts; that is, they thought that atoms and molecules were abstractions, and that those abstractions were predictive as analogies rather than actually being real things.

Well, it turned out that the evidence for the existence of atoms and molecules really entailed an application of probability. And one of the major players in that analysis was Albert Einstein. In 1905, Albert Einstein wrote three incredibly significant papers, the two most famous were on the topics of relativity and light quanta—the photoelectric effect. But his third paper was on Brownian motion; it was his theoretical work on Brownian motion that allowed experimentalists to do actual measurements that confirmed the reality of atoms and molecules.

Well, Brownian motion was discovered, first of all in the early 1800s by a botanist by the name of Robert Brown. It was in 1827 that he made some microscopic observations of grains of pollen on the surface of water and he noticed that these grains of pollen appeared to just constantly move in a jittery way on that surface; the motion seemed to just be random and it never seemed to slow down or to stop.

Well, in 1905, Einstein hypothesized that Brownian motion was caused by actual atoms and molecules hitting the grains of pollen, and thereby impelling those grains of pollen to take this random walk on the surface of a liquid. And remember, the random walks were what we talked about in the last lecture—randomly moving from place to place in different directions by this random motion of the molecules.

Well, Einstein wrote down a formula that basically predicted what distance a piece of pollen would move on average per unit time; that is, how far away it would be displaced from where it started per unit time—the kind of analysis we talked about in the last lecture, how far away you move—and then his formula actually gave a quantitative measurement to what you would expect. Well, anyway, his formula involved the size of the grains of pollen and the temperature and the viscosity of the liquid, and his formula predicted visible effects of things that were invisible; namely, the atoms and the molecules. They were too small to see, but they had an effect that was visible on these grains of pollen. He predicted that on average a pollen grain would be displaced from where it was at a given moment at a rate proportional to the square root of time. And this prediction was a probabilistic consequence of the hypothesis that atoms and molecules really existed and that their motion was just random. Well, this analysis then created the opportunity for the theory of atoms to be confirmed experimentally. And the experiments were actually undertaken in 1908 and beyond, and since the experiments accorded with Einstein's predictions, the experiments were strong evidence for the actual existence of atoms.

In a sense, Einstein's work encouraged a mode of reasoning that led later to the inherently probabilistic nature that's involved in quantum mechanics. See, in quantum mechanics, the most fundamental objects that make up matter are to be considered not as little round balls that are at one place at one time, but instead they're to be viewed as a probability distribution; they have a probability of being anywhere at any time. And this concept of the probabilistic nature of matter that we find in quantum mechanics is a real conceptual challenge.

Einstein himself never accepted this probabilistic nature at the root of quantum mechanics. He had a very famous quote that expressed his opposition to this probabilistic view. He said, "God does not play dice with the universe," to which Niels Bohr replied, "Einstein, who are you to tell God what to do?" And in fact later Stephen Hawking added, in talking about black holes: "God not only plays dice, but sometimes throws them where they cannot be seen."

So, in any case, probability is certainly involved in descriptions of physics. At the chemical level, chemical interactions really are to be viewed as the vast quantities of molecules and liquids or gases moving randomly around and encountering each other at random moments to create the products of chemical reactions. The fact that there are so incredibly vastly many particles tells us that the expectations about what we predict probabilistically can really be completely relied upon, because ultimately we're talking about the law of large numbers—that if we have so many experiments going on, the thing that we predict is certainly going to happen.

Probability plays a central role in physical theories from quantum mechanics up the ladder of different sizes of interacting matter up through chemistry, and then all the way up to macroscopic matters such as the weather. So now we're going to turn our attention to the discussion of the most familiar experience that we have on a day-to-day basis about probability, which is reading in the newspaper every day the statement: There is a 30% chance of rain tomorrow.

What we're going to do is now turn our attention to the question of what in the world does that statement mean? What's surprising is that that statement is a very difficult thing to understand, and if you ask a lot of people you'll discover that they don't really understand what it means to say there's a 30% chance of rain tomorrow. So we're going to start to think about this by thinking about what we want it to mean, and then seeing what it actually means, and seeing whether or not those agree.

Okay, so let's first of all take care of a couple of minor issues that are easily resolved. The first issue has to do with the issue of threshold, meaning that, how much rain is rain? Well, that's just decided by choosing an arbitrary amount. We say that rain consists of, it's usually chosen to be like 100th of an inch; if there's 100th of an inch, then that counts as rain. So the first thing is this threshold issue.

The second case, that's actually simple, so we'll do this second case first, is what does it mean to say that there's a 30% chance of rain at one specific spot? At one specific spot, what's the chance of a 30% chance of rain? Well, that means what you would probably guess: Namely, that if the weather circumstances are like they are today, then in 30 out of 100 subsequent days, if you're predicting the next day's weather, in 30 out of 100 of those subsequent days, you would expect to find at least 100th of an inch of rain landing at that spot. That's what it would mean to say there's a 30% chance of rain at one individual spot.

Now, let's be clear. It does not mean that it's expected to rain for 30% of the time. That's not true. It just means that if the conditions are like today, then tomorrow, if we're predicting the weather for tomorrow, there's a 30 out of 100 days—30% chance—that it will rain at least a measurable amount at that spot.

The issue comes in of what happens when we're talking about a whole region. When we say there's a 30% chance of rain in a region, what does it mean? The reason that it's a challenge is that there are different points in the region. Maybe the region contains a hill where it rains more often than it does in the valley area where it rains less often. So we have to understand how we're going to deal with the fact that there's some variation in the probability of its raining at one point versus another.

Well, let's start thinking about this issue by looking at the simplest case. The simplest case is if at every point in the region it really did have exactly the same probability of rain. In other words, suppose that the region that we're talking about—maybe it's either very small or very homogeneous in its character—that it literally were the case that at every single individual point, the conditions today predict that at each of those individual points there's a 30% chance of rain at each point.

Here I have a graphic that will show you this situation. If the area is homogeneous—that is to say that at every individual point in the whole area, there is a 30% probability of rain—then we'll summarize that by saying there's a 30% probability of rain in the area. Now, I hope that that is natural, and it's correct, and we all agree.

The challenge comes when parts of the area have a different probability of rain, point by point, than other parts. Let's look at this example here. Suppose that we have a 100-acre tract, and at half of that area, 50 acres, at every point in that region, those points have a 40% probability of rain for the period that we're talking about—tomorrow, from 12 midnight to 12 midnight the next day, there's a 40% chance of rain at each of those points; but then at the other half, the other 50 acres, at each point in those, there's only a 20% chance of rain. If you want to envision a mountain area and then a valley area, where more rain comes and it's twice as likely to rain in the mountain slope than on the valley, then at each point of the valley there's a 20% chance of rain.

Now, since we're trying to give useful information to everybody in the region, we can't go point by point, and tell each person in the region what the probability of rain is at your specific point. We need to summarize this, and we're accustomed to doing this. We know how to summarize it. The way we summarize it is we say, "Well, on average…"—it's taking an average of what the probability of rain is over the region.

One way to phrase the taking of the average is to say it the following way. Suppose we choose a point at random in the whole region, what is the probability of rain at that random point? If we phrase it in that way it becomes an expected value computation, because, you see, we could choose a point that's in the 40% area or we could choose a point that's in the 20% probability area. And there's a 50% chance, because half the areas are each, there's a 50% chance that we'll be in the 40% probability part, and there's a 50% chance we'll be in the 20% probability part. So on average, the expected value of the probability of rain over all the points in the region is 0.5 × 40% probability + 0.5 × 20% probability, which is the average; namely, 30% probability of rain.

Let me talk about an analogy of the kind of thing that we're doing. Suppose we had a roomful of people, and we wanted to give one number that summarized the height of the people in the room. Well, if everybody were exactly the same height, our summary would be simple. We'd say the height of the people in the room is whatever that height is—5'6". If everybody is 5'6", no problem; but suppose that some people are tall and some people are shorter. How would we give one number that summarizes that? Well, we'd have to give an average, some sort of a way to average it.

And one way we could say it is, if we choose a person at random in the room, what is their height? Well, what does it mean to choose a person at random? What it would mean is that if you have, say, two people in the room, and one person is 6' tall and one person is 5' tall, then you can't choose either of those heights. You would choose 5'6" as the average height. If you randomly chose somebody in the room, you'd either choose the five-footer or the six-footer. Neither of them would be correct, but on average, if you chose a lot of times, the average of the heights that you chose would indeed be 5'6", which is your prediction. Okay, so that is what we're going to mean by the probability of precipitation.

Let's give some more examples. Here's an example: suppose that we had two parts of our 100 acre region, 30 acres of which definitely would get rain. At every point it's definitely going to rain. It rains every single day at that whole 30 acres, and at the other 70 acres it never rains at all. Now, it's too bad that we have to give one number to summarize the probability of precipitation for tomorrow in this situation, but if we have to do it, how would we give a summary? We would say, "There is a 30% chance of rain," because if we choose a point at random in that 100 acres there's a 30% chance— 0.3 chance—that there's 100% chance of rain, and there's a 70% chance that it's not going to rain at all—that there's a 0 chance of rain. So the expected value of the probability of rain over all the points randomly selected in the region is 30%. So this is the way that we're talking about how we summarize what it means to say the probability of precipitation in a whole region is 30%.

Now here are some other examples, and of course you can think of all sorts of examples. Here's one where a quarter of the region, 25 acres, has a 40%; 50 acres has 30% at each point, and 25 acres, 25%, has a 20% chance of rain. Once again, the expected value is 30%. So these are all just different examples of computing this expected value.

We actually now have the definition of what it means for the probability of precipitation to be 30% for this defined period tomorrow. There's a consequence to this definition, and the consequence is the following statement: That on average, the average amount of the area that will get rain is 30% of the area—on average. Now, let me be very clear on what that means, to say that on average 30% of the region will get rain on a day in which the probability of precipitation is 30%. What it means is the following. Suppose that you imagine the rain happening many, many days. In other words, you have this prediction of a 30% chance of rain, and then you have that exact same weather condition for many, many days; and to be simple, let's say 10 days in a row. Ten days happens where exactly the same 30% probability of precipitation comes about, and every day a certain fraction of your area gets rain. Some days none of it rains—it doesn't rain at all; some days it may rain over the entire area; some days it rains over half the area; some days it rains over a third of the area. If we took the total of all of the amount of area that got rain during those 10 days, and then divided by 10, that's what we mean by the average area that gets rain per day. Does that make sense? I hope that makes sense. That's what we mean by the average.

Here we go. Let's do some examples. The easiest example is this case. Suppose we're in the case where we have our region in which 30 of the 100 acres always gets rain, and the other 70 acres never gets rain. Well, then, in fact, every single day 30% of the region gets rain, because those 30 acres get rain, 70 acres do not get rain. It happens every day, so on average it happens that 30% of the region gets rain.

Let's consider where the probability of rain at every single point is 30%. The probability of rain at every single point is 30%. Just to help with the graphics, let's assume our region is really, really thin, so that I can draw several copies of region on the screen at one time. Here's this long, thin region, and I'll break it down into a lot of little pieces, and for convenience here, I've just broken it into 10 pieces, but you can in your mind think of breaking it into like little square inches and have millions of them. But here we'll just assume that we have 10 little pieces, and these are so little that either the whole piece gets rain or the whole piece does not get rain on any given day.

Now, suppose that at every point in this region there's a 30% chance of rain. Now what does it mean at each point over 10 days? What will happen? Well if the probability at that point really works out, and let's assume that it does, at every point your prediction is accurate; and if you'd like to, you can think instead of 10 days that we're doing thousands of days, and so by the law of large numbers we expect 30% of them to get rain. But we'll just assume it happens in 10 days, so that means that this point—and I'm going to draw the same region in 10 different lines here—and looking at one point, I have that point, and I imagine 10 future days, and at 3 of those days it rains, and notice I've drawn them blue, meaning that it got wet.

That means that this particular sub-area of our region, this little tiny part, on three of our hypothesized 10 days in which this 30% weather prediction accrued, that on 3 of those days it got wet. Then on the other 7 days it did not. And we assume that that happened at every single point—you can see that if I look at a point on my region, my region's just this thin region, and I'm imagining 10 future days after our prediction—and if I look at every vertical line, I'm saying, "Okay, I have this point of the region," and I'm saying, "How many days did it get wet?" You'll notice that I've carefully drawn it so that every vertical bar has three blue days, meaning that it rained 3 out of the 10 days—on every blue bar.

One day, day four here, it rained every single place. In other places, like here, it just rained in 2 of our segments, but every vertical bar has 3, because we're assuming in this case that at each point the probability of rain is 30% and that over these 10 days that each point does get that probability of rain.

What is the average rain over those 10 days? Well, the average amount of area that gets rain over 10 days is, we just sum up the total amount of area that got rain, and divide by 10. Well, since every vertical line here has 3 blue days, then we have 3 times (and we've divided our area into 10), so we have 30 blue segments, and we divide those 30 blue segments by 10 days to get an average of 3 blue segments per day out of a total of 10 segments in the region. That is an average of 30% of the region getting rain per day. So this is the fact: That if you have 30% probability of precipitation, then on average over time, 30% of the region will get rain. And by the way, the same thing works if we did this example where half the region got 40% and half the region had a 20% chance of rain per region. The same kind of analysis would show you that the number of blue squares that you fill in, if every vertical column accords with its probability—that is, on this side 4 of the vertical things are blue, and on this side 2—that once again you'd get the same number of blue squares in this whole region—30 out of 100—and so that means that 30% of the region gets rain *on average*.

We've now done two things: we've showed the actual definition of what it means to say there's a 30% chance of rain tomorrow in a region, and we've showed that one of the consequences of that is that on average 30% of the region will get rain—on average over many days, if the probability of precipitation is 30%.

This is a very tricky definition, and many people misunderstand it; that is, the definition of probability of precipitation. It's tricky because what we're doing is we're taking an expected value of probabilities. We're averaging over different probabilities over different places. It really is sort of tricky. And one thing that makes it more tricky is that the official definition is at best misleading, and, I would argue, actually wrong.

Here is the National Weather Service definition of the probability of precipitation. It says:

> Technically, the probability of precipitation…is defined as the likelihood of occurrence (expressed as a percent) of a measurable amount (.01 inch or more) of liquid precipitation (or the water equivalent of frozen precipitation) during a specified period of time at any given point in the forecast area.

And then it goes on, "Forecasts are normally issued for 12-hour time periods." The salient phrase is "at any given point in the forecast area."

This definition cannot be true, because if your area has one point in the mountains where there's a higher percentage probability of precipitation than another point in the valley where it has a lower percent probability of precipitation, it cannot be true that the single number gives you the likelihood at any given point.

Now let me be very clear: The meteorologists at the National Weather Service are completely clear on the correct definition of probability of precipitation. They don't have any misunderstanding. The problem is the phrasing of this definition. They don't mean "at any given point." What they mean is that at any *random* point—at a random point in the region—but then when you say a random point, you don't mean at any given point. You mean that you choose the point randomly, and then take an average.

Well, people have written papers, including modern papers written in just the last couple of years, about the fact that the public misunderstands the definition of probability of precipitation. But what was disturbing to me, is that apparently the people who wrote these papers weren't so clear about the definition, either, because here's an example of a multiple choice test that was given to people to determine how well they understand the phrase that "the chance of rain is 30%."

Here are the choices:

    (a)   Rain will occur 30% of the day.

(b)   At a specific point in the forecast area, for example, your house, there is a 30% chance of rain occurring.

(c)   There is a 30% chance that rain will occur somewhere in the forecast area during the day.

(d)   Thirty percent of the forecast area will receive rain, and 70% will not.

Now notice that none of these is correct. Rain will occur 30% of the day—we dispensed with that. That was not correct. (b) is the one that in the paper they said was the correct choice to make; but it's not correct, because you see, they misinterpreted the phrasing in the definition that said "at any given point" to say, for example, your house. But that is not true, because your house may be one place, some other place is a different place, and you have different probabilities of precipitation. It is simply false that it's saying that "At a specific point in the forecast area, for example, your house..." That's just not true.

Now, these other ones aren't correct either. "There's a 30% chance that rain will occur somewhere in the forecast area during the day." That would be a possible definition for 30% chance of rain in a region, but that wouldn't be a definition that you would want to make. The reason that you wouldn't want to make that as a definition is, suppose you had one place in your forecast area where it always rained, you know, somehow the plants always caused a little dew to come up and it rained a little tiny bit every single day; but most of the region had some variability. You wouldn't want to say there's 100% chance of rain because that one little part there was a good chance of rain. So that wouldn't be representative of the whole area, it wouldn't give the information that you want to give if you're talking about a whole region. So (c) is not correct.

(d) "Thirty percent of the forecast area will receive rain, and 70% will not." Now actually (d) would be correct if instead of saying "will" they would say "on average will" receive. You see, that's what we proved—that on average it is correct that 30% of the forecast area, on average, will receive rain, and 70% will not. However, if you remove probability, and say that 30% actually will receive it, that is certainly not true, because the prediction of rain is an inherently probabilistic matter. People don't know whether it's going to rain or not. So a probability is involved, so you can never interpret it as saying "will."

Well, I found this topic to be a fascinating example of where a rather technical issue associated with probability can really trip up people, and for good reason, when we read things every day in the newspaper. What does it mean for there to be a 30% chance of rain? You have no idea how many hours I have spent talking to friends about this issue, and it's really, really fascinating.

But I'll make you a bet right now, since we're in the probability game, I will bet $1 that within five years the phrasing of the definition by the National Weather Service will change, because somebody will hear this lecture and go ahead and change the definition of what it means to say there's a 30% probability of precipitation. We'll see.

I'll look forward to seeing you next time when we're going to talk about probability as it applies to biological issues. I'll see you then.

Lecture Six
Probability Is in Our Genes

Scope:

One of the most basic issues in biology is to describe how characteristics of parents are passed on to their offspring. The basic idea is that each parent randomly contributes part of that parent's genetic material to the offspring. The combination of genetic material received from the parents determines characteristics of the offspring. Because randomness is centrally involved in the passing down of genetic material, genetics, the science of inheritance of traits and characteristics, is modeled probabilistically. The simple Mendelian model of dominant and recessive genes provides a probabilistic answer to the question: What traits will the offspring of two specific parents have? Then, probability is used to show the distribution of traits over a whole population and to describe how the characteristics of the whole population will alter through a random process called *genetic drift*. Probability lies at the very core of biological descriptions of mutation and evolution.

Outline

I. Genetics, the science of inheritance of traits and characteristics, is modeled probabilistically. This lecture discusses three probabilistic aspects: the Mendelian model of genetics, genetic drift, and mutation and evolution.

II. The simple Mendelian model of dominant and recessive genes is the basic model of inheritance. For the sake of simplicity, we will use brown and blue eye color to illustrate this concept, and we will make the simplifying assumptions (though they are not true for real people) that a single gene determines eye color and that there are only two possible colors, blue and brown.

   A. The Mendelian model gives a probabilistic answer to the question: What traits will the offspring of two specific parents have?

   B. Different versions of a given gene are called *alleles*. In our example, these would be brown (B) and blue (b). People will have BB alleles, Bb alleles, or bb alleles.

C. Each parent contributes one allele for a given gene, either B or b.

D. If either of the alleles in the offspring is the dominant type (B), its trait will be expressed. Otherwise, the recessive trait (b) is expressed.

E. Therefore, the probability of the recessive trait (b) being expressed is $\frac{1}{4}$ if both parents carry one recessive allele, as shown in the chart that follows.

Parent	B	b
B	BB	Bb
b	Bb	bb

F. The chart below shows the percentage breakdown of the offspring (in the shaded area) if we imagine that 60% of the alleles in the parent population are for brown eyes and 40% are for blue eyes.

Parent Alleles	B 60%	b 40%
B 60%	BB 36%	Bb 24%
b 40%	Bb 24%	bb 16%

G. If we imagine a representative population of 100 offspring, each with two alleles (BB, Bb, or bb), note that the proportion of brown to blue alleles has not changed from the original:

36B + 36B + 24B + 24B = 120 B alleles (60% of 200)

24b + 24b + 16b +16b= 80 b alleles (40% of 200)

H. The *Hardy-Weinberg equilibrium theorem* shows that even if you have a recessive characteristic, it will not disappear. Instead, there is a stable percentage that remains as generations pass.

I. The Hardy-Weinberg equilibrium theorem applies to recessive disorders as long as those disorders do not have an impact on reproductive success.

III. Probability plays a central role in viewing genetics over the time scale of tens of thousands of years. Genetic drift alters the percentage of alleles that are dominant for a given trait.

    A. By random chance, the percentage of dominant alleles in the next generation is different.

        1. The expected value of the percentage in the next generation is the same as the percentage in the present generation.

        2. But the actual percentage is often a bit different by chance, as our simulations show.

    B. This changing percentage is called *genetic drift*, and it can be modeled using the idea of a random walk.

    C. Genetic drift is most prominent when the population is small. It happens much more slowly in larger populations.

    D. All of this assumes that no natural selection is going on that affects the proportion of the allele.

        1. In other words, no trait has an advantage in the number of offspring that a person with that trait can reproduce.

        2. If such an advantage exists—for example, if each blue-eyed parent has an extra child, that selective advantage quickly takes over.

IV. Another way that genetic material changes is through mutations.

    A. A mutation is a stable change in the genetic material, brought about by various means, transmitted to offspring.

    B. Mutations to nonessential portions of the DNA are useful for measuring time (the molecular clock).

        1. It is assumed that mutations to nonessential aspects occur with a uniform probability per unit of time in a particular portion of the DNA.

        2. If P is the probability that a single segment of nonessential DNA has no mutations in a year, then P^Y is the probability of no mutations in a segment of DNA happening over Y years.

3. On the average, if you have two individuals who had a common ancestor many generations ago, you would expect them to have about the same percentage P^Y of segments of nonessential DNA that had no mutations.
4. Assuming that mutations are so rare that it is very unlikely that a mutation in the same segment has occurred in the two individuals, the percentage of segments that are mutated in one or the other is, on average, $2(1 - P^Y)$. This is an estimate of the percentage of segments that would be found different if comparing two individuals with a common ancestor Y years ago.
5. Using this kind of probabilistic inference, we can estimate that the most recent common female ancestor of all living humans lived about 150,000 years ago.

V. Let's look at a hypothetical situation that has a probabilistic aspect: universal HIV testing.

A. About 1% of the time, HIV tests give a false-positive result.

B. Of those who have HIV, their tests will come out positive 95% of the time.

C. If someone has a positive result, what is the probability that that person has HIV? Let's look at the numbers:
1. Let's say the population of the United States is about 300,000,000, of which about 500,000 people are HIV-positive.
2. Of the 500,000 who actually have the disease, the test will come out positive 95% of the time, which equals 475,000 cases.
3. There are 299,500,000 (that is, 300,000,000 − 500,000) people who do not have the disease.
4. Of the 299,500,000 people who do not have the disease, the test will come out falsely positive 1% of the time, which equals 2,995,000 cases.
5. Thus, the total number of people receiving a positive test result is: $475,000 + 2,995,000 = 3,470,000$.
6. But of the 3,470,000 who get positive test results, only 475,000 actually have the disease. Therefore, if you get a positive test result, your probability of having the disease is $\dfrac{475,000}{3,470,000}$, which is less than 15%. This is an

example of a probabilistic anomaly that is an artifact of giving universal testing for a rare disease when the tests have a significant possibility of giving false-positive results.

Readings:

Brian Charlesworth and Deborah Charlesworth, *Evolution: A Very Short Introduction*.

Questions to Consider:

1. Assume that for some gene, there are more dominant alleles than recessive alleles in the current population. How can you reconcile the following facts: First, that the expected value for the percentage of a recessive allele in the next generation's population is its current percentage, and second, the percentage of that allele is probabilistically expected to become only half as great as it is now or twice as great as it is now at some point in the future?

2. How could the rate of change in nonessential parts of DNA be used to disprove the theory of evolution if it were false?

Lecture Six—Transcript
Probability Is in Our Genes

Welcome back. Today's lecture is going to be about probability as it applies to biology.

One of the most basic issues in the biological sciences is to describe how the characteristics of parents are passed on to their offspring. The basic concept of genetics is that the genetic material from each of the parents is randomly combined; that is, part of the genetic material of the father and part of the genetic material of the mother are combined to become the genetic material for the offspring. And the offspring then have different traits according to which material was contributed by the two parents.

Since the contributions from the parents have a random characteristic to them, the concept of probability plays a central role in how to predict the characteristics of offspring. In other words the genetics, which is the science of inherited characteristics, genetics gives probabilistic answers to questions such as: How will the offspring be in relationship to how the parents are?

In this lecture we're going to be talking about three things. One is we're going to be talking about the Mendelian model of genetics. We're going to be talking about, then, genetic drift, and talking about mutations and evolution.

Let's begin by just recalling the basics of Mendelian genetics, which you all know from high school days, but let me just describe it in terms of brown eyes and blue eyes. Now, of course, I know that the fact is that the characteristic of eye color is not determined by a single gene, but for purposes of this lecture, which is about the probability associated with genetics, it will be. We're going to just say that a single gene determines eye color—it doesn't, but let's just assume that it does.

And every gene, such as the one that's going to determine the eye color, has several different possible variations to that kind of gene, which are called the *alleles*. And so you could have a brown allele or a blue allele, and each individual, each person, has two alleles that make up their gene of the eye color gene, and they could have any of the following types: they could have brown-brown alleles; they could have a brown-blue alleles; or they could have blue-blue alleles.

So that's the basic genetic makeup for each individual under this, again, hypothesis that there's a single gene that determines eye color.

Then there's the concept of dominant versus recessive. Remember that if either of the alleles in a person's genetic makeup for eye color is brown, brown being the dominant allele will express itself and the person's actual eye color will be brown. So in our hypothetical world, by the way, everybody has either brown eyes or blue eyes, and the genetic makeup of each individual in that gene consists of either a brown-brown, brown-blue, or blue-blue combination.

When we talk about offspring, here is the kind of a chart that you often see associated with how to determine what the allele makeup will be for the offspring. Suppose that you have a parent that has a brown allele and a blue allele, and the other parent also has a brown allele and a blue allele in the eye color gene. Then each parent randomly contributes either the brown or the blue allele. If this parent randomly contributed its brown allele, and this one contributed its brown allele, then we would have two brown alleles as the genetic makeup for the offspring.

If this parent contributed a blue allele, and this parent contributed a brown allele, then the offspring would have a brown and a blue allele, but the eye color would be brown, because having a brown allele, and brown being dominant, means that the eye color is actually brown. Likewise this quadrant refers to this parent's contributing a blue allele while this one contributes a brown allele—eye color is brown—and only in this quadrant will both of the alleles be blue, and consequently the eye color actually will be blue.

So if one has two parents, each of whom has both of the alleles, then the probability of having a child with blue eyes would be 1 out of 4. Okay, so this is basic genetics, and I think you understand this.

Let's look at it slightly differently, however, and look at it from the point of view of the whole population. Instead of thinking about two individuals, suppose we have an entire population of people. And these people, some of them have brown-brown alleles for their eye color, some have brown-blue, and some have the two blue alleles, and are blue-eyed. Let's just imagine that in the whole population, 60% of the alleles—not the brown color, but the alleles—are the brown alleles, and 40% of the alleles in the population are the blue alleles.

There are many ways that this could happen. For example, it could be that 60% of the people have brown-brown alleles, and the other 40% have blue-blue alleles. And that's the way it's distributed. But what we're going to think about now is suppose that randomly we select a mother and a father, and they randomly come together and mate to have offspring. What percentage of the offspring will have the different eye characteristics and the different allele characteristics? So we're looking at it probabilistically in the whole population, rather than looking at just two individuals.

Under the assumption that there are 60% brown alleles in the population, then the probability of choosing a brown allele from the male in the population is 60% (0.6). The probability of choosing a brown allele from the female population is also 0.6. So the way to think about it is that if you have the collection of females in the population, each female has two alleles. Well, if you randomly choose a female, and then randomly choose which of the two alleles is going to be contributed, every allele in the whole female population has an equal chance of being selected, so that's why we can say there's a 60% chance of choosing a brown allele from the female population, and a 60% chance of choosing a brown allele from the male population.

Here we have, then, the probability of getting a brown from the male and a brown from the female is the product of (0.6 × 0.6) which is 0.36 or 36%. That means that randomly choosing an allele from the males and an allele from the females, if there are 60% brown alleles in the males and 60% brown alleles in the females, then the predicted percentage is 36% of brown-brown allele combinations among the offspring.

Likewise, the same kind of computation, there's a 60% chance of getting a brown allele from the male population—this column is the males—and a 40% chance of getting a blue allele from the females, so the product from those two: (0.6 × 0.4) is 0.24, so 24% of the population will have brown-blue alleles by virtue of getting a brown allele from the male contributor and a blue allele from the female contributor. Likewise, there'll be 24% that also have brown-blue, but this time the brown from the female and blue from the male contributor; and 16%, that is (0.4 × 0.4), is the probability that we'll choose a blue allele from the male population and from the female population. So that means that 16% of the population will have blue eyes.

Well, notice that this collection of probabilities of these different proportions of the population that have the different combinations of alleles, notice that the proportion of the different alleles is exactly the same as it was before. For example, suppose that there are 100 individuals in the population, then that's a total of 200 alleles, because each individual has two alleles—well, there would be (36 + 36) brown alleles, there would be 24 more brown alleles here, 24 more brown alleles here, and the total of (72 + 24 + 24) is 120.

And the number of blue alleles: well, there are 24 blue alleles here, 24 blue alleles here, for a total of 48; plus (16 + 16) blue alleles, a total of 32 blue alleles here; for a total of 80 blue alleles. So the total number of alleles in the offspring population is 120 of the brown alleles, and 80 of the blue alleles, which you notice is in the same proportion as where we started—60% are the brown and 40% are the blue.

What this says—and this is by the way called the Hardy–Weinberg equilibrium theorem—it says that if we have a certain proportion of blue and brown alleles in the adult population, and we randomly mix the parents, then the offspring will have exactly the same proportion of blue and brown alleles. What this means is that even if you have a recessive characteristic such as blue eyes, you might think that as generations pass the percentage of blue eyes would sort of die out, but in fact the Hardy–Weinberg consequence is that there's a stable percentage of those recessive alleles.

What that means is that diseases, for example, that are rather rare, like the cystic fibrosis disease, that we expect that on average the percentage of people who have that disease will be exactly the same over time. Of course, it'll die out if there's a smaller chance of reproduction, but we'll get to that later. At this stage we're just saying, assuming that they have the same percentage of probability of reproduction, then the percentage of alleles, dominant or recessive, will tend to just stay the same over time; that is the consequence of the Hardy–Weinberg equilibrium theorem.

So at this stage, what I have tried to demonstrate is that there is a constancy in the percentage of dominant and recessive alleles. What I will now say is that we expect the percentage of dominant and recessive alleles to change; in other words, exactly the opposite—and the reason is that probability is involved. So although the Hardy–Weinberg equilibrium theorem tells us that on average that if the expected value of the percentage of dominant and recessive alleles remains constant over time, in reality, since probability is involved, there is a chance that the probabilities will shift a little bit.

Well, in fact, the shifting is like a random walk, you see, because when you have an actual population that randomly comes together, it won't be the case that the percentage of offspring exactly come out the way that the expected value tells us that they should. Instead, they'll be off a little bit. In other words, they've drifted a little bit. And then, the next step, they'll drift a little more.

Well, here we've done some simulations to demonstrate this. Suppose that we start out with a population as we described before. In other words, it has a population where 16% of the population is blue-eyed, and 40% of the alleles are blue alleles, as we saw in the previous slide. That means 16% of the population actually have blue eyes, because they have both blue alleles. And then 60% of the alleles in the population are brown-eyed alleles.

Now what we're going to do is do some simulations where all we do is take that situation and have random generation of children. Here we have a case where we start out where there are 16% of the population blue-eyed, and we randomly have them combine to make the next generation. And we're thinking of an extremely small population, that is, about 25 individuals in the whole population, and look what happens. In the first generation, instead of having just 16%—it started with 16%—but it went up to a higher percent, maybe 18%. Then it went down, dipped below where it started; it went down again; went up a little bit, and you can see after a few more generations it has risen to be now more than 20% of the population is blue-eyed. Now it dips down again; dips down and up, and you can see it's taking sort of a random walk, varying up and down by chance alone, and then after a certain number of generations, in this case about 50 generations, it dies out and there are no more blue-eyed people.

Now, notice that this simulation has nothing to do with natural selection or survival of the fittest; we'll talk about that later. But what this is saying is, just by random chance alone, the random walk fluctuations of the percentage of blue-eyed people is expected to change by just randomness—and in this case it died out.

Let's just look at some other simulations of this same feature and see what happens. Here we've done the same simulation again, but we'll do 20 such simulations and see what happens. In this simulation, sure enough, the blue-eyed people die out after a few generations; this simulation they die out again; this simulation they last for about almost 100 generations before they die out; but look what happens in this one. In this one, just by random chance alone, the blue-eyed people take over, and there are no more brown-eyed people. We have, by random fluctuation and this random genetic drift—which is like a random walk—here's another example where the blue-eyed people gained in population and pretty soon took over the entire world. Here they died out. So in our simulations we can see this variety of sometimes coming to dominate the whole population and sometimes dying out.

Now, one of the reasons for the swiftness of these changes is that we're talking about a very small population; namely, we're talking about just 50 alleles in the whole population—a population of about 25. And the genetic drift happens much more slowly when we're talking about larger populations.

Let's do some simulations of larger populations. Suppose you have a population of 2,000 people, for example, but we have the same proportion of alleles; that is to say, 60% brown-eyed alleles and 40% blue alleles, so that 16% of the population is blue-eyed. Now we're going to do the same kind of simulation again, and doing these kinds of simulations where we show the percentage of blue-eyed people— that's the blue part of the graph—and then the brown graph is the part of the population that is homozygous brown, meaning that both of their alleles are the brown alleles.

Notice in this simulation the blue population, the blue-eyed people, actually became more numerous than the homozygous brown-eyed people for a while, and then dipped down, and in fact at the end, after 10,000 generations, there were more blue-eyed people than brown-eyed homozygous people.

Now, of course, everybody who is not blue-eyed is brown-eyed, but more than the percentage of those who are homozygous brown-eyed. Remember, the percentage of people who are homozygous brown-eyed were 36% to start with, because it was (0.6×0.6)—0.6 being the percentage of brown alleles in the population.

Here's an example where the blue-eyed people almost go to zero. Here more of the blue-eyed people do well; but notice this example, the brown-eyed people in fact become completely dominant. That is to say that there are no more blue alleles in the entire population. So, sometimes one of the alleles can actually take over by random chance alone.

Now, of course, one of the most basic parts of the strategy of genetic description has to do with natural selection; namely, that if a certain trait has an advantage in the number of offspring that a person with that trait can produce, then that trait becomes more dominant over time.

And in fact we can do some simulations to demonstrate how quickly such dominance prevails. For example, suppose that we assume that each blue-eyed parent—that is, a person who has both the blue alleles—has an extra child at each generation. Now of course, that's a huge advantage to think of one extra child when we're assuming that there's an average of two children per parent, so this would be a very strong competitive advantage in offspring if one were blue-eyed.

Under that assumption, which is the kind of assumption made in evolution, although not to that extremity, by the way, but the concept is the same—that certain traits have an advantage in reproduction— then look what happens to our charts. The blue alleles very quickly take over the entire population. In all of these simulations, that's what we would observe when we allow the blue-eyed people to have such an advantage in their offspring.

This is one of the features that makes natural selection and the theory of evolution work. Genetic drift, however, has nothing to do with natural selection. It's just a result of our understanding of what to expect from randomness.

Well, there are other ways in which genetic material changes. There are other ways in which genetic material changes, and that is through mutations. One way that genetic material would change is if you have a particular part of the gene—and it might be hit by a cosmic ray, for example, or maybe in the transcription of that genetic material as it's replicated maybe an error occurs—and you have a mutation.

Now, one can experimentally estimate the probability with which a single section of non-essential DNA mutates. There are some parts of DNA that don't seem to do much, and so those are good parts to observe to see the rate at which mutations happen; and so one can estimate the probability that such a segment of DNA will have a mutation. Then we can ask ourselves the question: For a given segment of DNA, what's the probability that after Y years it will never have had a mutation? So P is the probability that it does not mutate in one year, so $P \times P \times P \times$ itself the number of years is the probability that it will not mutate over Y years. So P^Y is the probability of having no mutations in this section of DNA for Y years.

Well, if we have many sections of DNA that we look at, and we say, "Okay, the probability of this section has P^Y of no mutation, and this one P^Y of no mutation, and this one P^Y of no mutation, then, like the law of large numbers, if you have a lot of sections, you expect that the fraction of the sections with no mutation to be equal to the probability of no mutation. In other words, we would expect that the fraction of the sections with no mutation would be about P^Y after Y years."

So on average, if you had two different individuals that started many generations ago and years past, and there are many generations that occurred, you would expect that two individuals who had a common ancestor many generations ago and then no further influence, you would expect them to have approximately the same number of segments that had no mutations in these segments of non-essential DNA. Well, you can make an estimate then for how many are different. I mean, how could you tell if they're different? Well, some of the segments are the same, so you would guess that those were not cases of common mutation, so you can estimate how many segments actually do observe mutation by comparing the different organisms, and taking that fraction of the sections that have differences in the two individuals, we can estimate that the average number of sections that are different should be counted to be about 2 × the probability of getting a different mutation after Y years.

Now, by the way, all of this is a very simplified version of what really happens. But the point I'm trying to make is only this: You can compute an expression that associates observations of DNA in individuals living today with a number that contains things that you know—the number P, which is something that you can deduce from observation—and the Y that you don't know. The y is when did these two individuals have a common ancestor.

What we can then deduce from this is we can figure out what Y is to make this equal the observed fraction of different segments of DNA that we observe in these two individuals. This kind of reasoning allows us to deduce when the most recent common ancestor of us all lived, and the answer is, doing this kind of probabilistic interpretation and inference, we can conclude that the common female ancestor of all living humans today lived about 150,000 years ago.

I've given a style of the argument, but of course the details are simplified. But it gives you a flavor of how we can make such a deduction as this number of years in the past where we had a common ancestor.

I want to conclude this lecture on a different note; namely, to talk about the question of universal testing for HIV. This is an issue that comes up and it has an interesting probabilistic aspect to it that I think we might want to discuss. If you give a test to an individual, there's a certain chance that that test is correct and a certain chance that it gives a false positive. Some of the tests that exist today have the property that about 1% of the tests give a false positive, meaning a person who doesn't have HIV would be tested as having HIV. And some of the tests are about 95% correct for people who have HIV; that is to say that a person who has HIV will get a positive test 95% of the time.

Well, if we do a little bit of arithmetic, we can ask ourselves the following question. Suppose that you went in to take this test, a random person goes in to take the test, and everybody in the country takes the test, and you got back a positive result. How would you feel? What's the probability that you would have HIV? Now, it sounds like what I've said is that you have a 99% chance of having HIV, because only 1% of the time when you have a person who fails to have HIV would the test come out positive. But let's look at the numbers, and the numbers tell a different story.

Suppose that we estimate the population of the United States at about 300 million, and of those about 500,000 people actually are HIV positive. Well, of the 500,000 who are HIV positive, 95%—which is 475,000—95% of the time the test will say that they have it. In other words, 475,000 tests would say correctly that that person has HIV. That means that there are 299,500,000 people who do not have HIV, but of those 1% will get a positive test result inaccurately. But 1% of this huge number, 299 million [sic 299,500,000], is 2,995,000 people will get a false positive on their test. So that means that the total number of positive tests that were received would be the 475,000 correct positive tests plus the 2,995,000 false positives, for a total of 3,470,000 positive tests. Of those 3,470,000 positive tests, only 475,000 people actually have the disease.

When you receive a card saying your test came out positive, in fact the probability that you actually have the disease is only 475,000 divided by the 3,470,000, which is less than 15%. So this is an example of a sort of probabilistic anomaly that is an artifact of giving universal testing where you have a rare disease and a significant possibility of getting a false positive.

In the next lecture we're going to be talking about applying probability to the world of finance. I'll see you then.

Lecture Seven
Options and Our Financial Future

Scope:

We've already discussed several applications of probability to gambling, and it seems natural that probability theory would arise in an area where great gambles are made—Wall Street. Predicting the future prices of stocks can have a significant impact on our view of our future financial security. Starting in 1900, a Frenchman, Louis Bachelier, devised the first model of stock prices that involved probability. We will also see that options contracts are fundamental to modern finance; in fact, more money is traded in options than in stocks.

Simply put, an option contract is an agreement between two people that gives one the right to buy or sell a stock at some future date for some preset price. Options are used as speculation, as well as a way to hedge risk, but it is a challenge to derive a rational price for such a contract. For quite some time, option pricing was viewed as a form of gambling. After the Black-Scholes theory was developed, the option price was viewed as an investment. As we will see from the example of Long-Term Capital Management, however, the application of sophisticated probability theory is not without its risks.

Outline

I. The world of finance is full of uncertainty, as is the world of gambling.

    A. Among many other financial issues, the future prices of stocks and options are definitely uncertain.

    B. If we want to evaluate whether our retirement fund is adequate, we need to consider what might happen to our investments and their values.

        1. We can take our financial portfolio and run probabilistic simulations.

        2. The probabilistic factors might include inflation or world events.

    C. Decisions about how much people are willing to pay for stocks are human decisions that are not predictable.

D. Randomness and probability play central roles in the determination of what our financial future is going to be.

II. How are prices of stocks or options modeled by financial mathematicians?

    A. In 1900, Louis Bachelier devised the first model of stock prices that involved probability.

        1. His model was basically a starting price plus a random walk.

        2. In Bachelier's model, the price varies purely randomly from its current price with equal likelihood of going upward or downward; underlying trends do not appear in the model.

    B. In reality, there may be some reason to believe that an asset will increase in value. For example, consider a cattle ranch that has lots of food and today has a small herd of cattle.

        1. We expect growth.

        2. The value of that asset will rise.

    C. Other assets, such as heating oil and corn, have cyclical trends.

    D. More robust, sophisticated models of future stock prices were developed that include a drift component. One model (Samuelson, 1960) incorporates three components: today's price, plus a function that relates to how the stock price is expected to change (the drift), plus a random walk feature (volatility).

III. Options contracts are fundamental to modern finance; in fact, more money is traded in options than in stocks.

    A. An option is a contract that gives the holder of the option certain specified rights.

        1. This might be the right to buy a security or a commodity at a specified price on a specified future date.

        2. Or it might be the right to sell a security or commodity at a specified price on a specified future date.

    B. We will talk about the simplest kind of options, namely, a piece of paper that says I can buy one share of XYZ stock on April 30 for $100, even if at that time, XYZ is trading for a higher price.

1. The possibility that XYZ will be worth more than $100 is what gives the option its value.
2. If XYZ is trading for less than $100 at that time, the option is worthless.

C. Options can be used as speculation and as a method to hedge risks.

1. Options used as speculation: If I contract the right to buy stock at a future time at $100, I am betting that the stock will actually exceed that price, so I can resell it at a profit.
2. Options used as a hedge against risks: Let's say I need copper for my business. I have a business plan, and I know I need a certain amount of copper at a certain price. I can buy an option to ensure that, at a future time, I can buy copper at today's price.

IV. How much should someone pay for an option? The idea of finding a rational price for options was developed in the late 1960s and early 1970s and allowed the options market to be created.

A. Let's take an example: I have bought an option that states that if XYZ stock, which now sells for $87, gets to $100 in the future, you pay me $1.

B. To determine how much I should pay to acquire that option, we can work out an expected-value analysis: If I believe the probability of the stock reaching $100 is 90%, then the option would be worth 90 cents. But someone else might feel that the option would be worth only 50 cents.

C. The rational price is one that enables the seller of the option to eliminate the risk and to ensure that he has the resources to pay out the $1 if the stock reaches $100.

1. If another person buys $\frac{1}{100}$ share today, then he owns $\frac{1}{100}$ of the stock.
2. And if the share reaches $100, the seller of the option can pay the $1 by selling the $\frac{1}{100}$ share.

3. Thus, the rational price for the option is the cost of $\dfrac{1}{100}$ share of our $87 XYZ stock today, or 87 cents.

V. Let us look at another example: Suppose an option is associated with a stock that today is selling for $100 per share, and we are talking about the option to buy a share at $100 one month from today.

   A. We make a simplifying assumption: The price will be either $110 or $95 one month from today.

   B. This concept of looking at a finite collection of possible future values at discrete moments of time is called the *Cox-Ross-Rubenstein* (*CRR*) *tree*. The CRR tree can be used to price options.

   C. Here, we try to replicate the risk of the option. We are going to buy a certain number of shares of stock and have a certain (negative) amount of cash in our portfolio. The value of our portfolio will be equal to the value of the option in one month's time. In other words, we are trying to quantify the risk itself. Here's the math:

   1. x = number of shares in the portfolio
   d = amount of cash in the portfolio
   If the price goes to $110, the option is worth $10.
   If the price goes to $95, the option is worth $0.
   $110x + d = 10$
   $95x + d = 0$

   2. Our solution is: $x = \dfrac{2}{3}$ and $d = \$-63\,\dfrac{1}{3}$.

   3. In other words, a portfolio containing $\dfrac{2}{3}$ share and owing $63.33 will have the same value as the option one month from now.

   4. Thus, the rational value of the option is the cost of $\dfrac{2}{3}$ share ($66.67) minus $63.33, or $3.33.

VI. This type of analysis leads to the *Black-Scholes model*.

   A. Before the Black-Scholes model, these contracts were viewed as a pure gamble.

B. The main result of the Black-Scholes theory is that the option price can be viewed as an investment, which led to the establishment of trading houses, such as the Chicago Board Options Exchange, created in 1973.

VII. The application of sophisticated probability theory is, however, not without its risks.

A. In 1994, the hedge fund Long-Term Capital Management (LTCM) began its historic money-making run, using advanced mathematics from top mathematicians.

  1. The man in charge was John Meriwether, a legendary head of bond trading of Salomon Brothers in the 1980s.

  2. He brought Myron Scholes and Robert Merton to serve on the Board of directors of LTCM. They later won the Nobel Prize in Economics for their work on options pricing.

B. LTCM used complicated mathematical strategies and sophisticated models to trade bond products.

C. In its first three years, to take full advantage of the bond mispricings their models found, LTCM borrowed heavily.

D. In 1998, LTCM collapsed. The Federal Reserve Bank of New York arranged a bailout of several billion dollars by 14 investment banks.

Readings:

Roger Lowenstein, *When Genius Failed.*

Questions to Consider:

1. Why do the prices of a given stock go down as well as up even when the company is doing well?

2. The future prices of stocks are uncertain. What option and stock portfolio could you purchase to guarantee that you will not lose more money than the price of the option even if the stock price falls dramatically, yet you still reap the benefits of substantial gains in the price of the stock? This is an example in which options are used to hedge against stock decline.

Lecture Seven—Transcript
Options and Our Financial Future

In previous lectures we discussed several examples where we applied the concepts of probability to gambling, so it seems only natural that we should turn our attention to how the theory of probability is going to arise in an area where really great gambles are made: namely, Wall Street.

Predicting the future prices of stocks can have a significant impact on how we view our whole future financial security. The question is: How are we going to model the behavior of stocks or other financial instruments so that we can have a guess as to whether or not our retirement fund is going to be adequate to keep us living in the lap of luxury? Or how are we going to view what will happen to our investments and their value over time? So we need to somehow think about how to guess what's going to happen in a future that we know is uncertain.

Well, one possible strategy is that we can just take our financial portfolio, for example, and just run simulations where we put into those simulations various probabilistic components. Now the reason that we have probabilistic components in them is because there are various aspects of what's going to happen to the future value of our investments that are not really determined at this time. Our financial future is going to be determined by such things as whether or not inflation happens, or whether or not a political event might occur that would have an effect on the value of our stocks. In other words, the future value of stocks isn't a completely deterministic kind of a study.

Many of the aspects that determine the prices of future stocks are things that are out of our control and haven't really been decided yet. One of the main reasons that we don't know what the future values of investments are going to be is that the decisions about how individuals, human beings, evaluate the price of a stock is a human decision, and human decisions are notoriously unpredictable. Randomness and probability play central roles in the determination of what our financial future is going to be, and that's why we view it as an example of a probabilistic kind of a scenario.

The question that we're going to be talking about during this lecture is: How do we build mathematical models for the case of trying to describe the future prices of financial instruments—such as stocks or options that we'll talk about in a minute? Our goal is to make these mathematical models about how things will behave in order for us to make rational decisions about how we should invest our money.

Well, back in the beginning of the 1900s, we had the first example of a mathematical model of stock price movement, and there was a gentleman by the name of Louis Bachelier who produced some models about stock prices, and his model had the following feature. He simply imagined that the model started with the price of the stock as it is today, and then the future price was determined simply by a random walk. In other words, it would fluctuate up and down, and up and down, according to just random fluctuation, but it wouldn't have any global trend.

Here we have some examples of some simulations of such price fluctuations based on just a random walk. So this is an example where the price started at this stage, and you can see that it randomly oscillates up a little, fluctuates a little up, a little down, a little up, and a little down. This would be a typical example of the price future of a stock under Bachelier's model.

However, Bachelier's model lacked something, and what it lacked is that there was no underlying trend associated with the value of a stock. This is something of a defect, because one can have the expectation that a certain kind of an asset or a stock will actually increase in value over time; that is, there's an underlying reason why we expect an asset to increase in time.

For example, let's take a cattle ranch. Here we're imagining investing in a cattle ranch; and the cattle ranch consists of some luscious land full of greenery, and then it also has a certain small herd of cattle who are enjoying themselves, and over time, we have the expectation that more cattle will live on this land, and the value of that asset will become greater. So there's an underlying trend that we expect. Now, of course, we don't know what's actually going to happen, and that's why we don't price the value of the cattle ranch according to what our expectation will be in 10 years. It will just gradually rise up as the cattle do in fact increase in number, as long as that actually happens.

That's an example of a case where a certain stock has an expectation for increasing in value at a rate that we might be able to predict by some external reason. Other kinds of assets have a cyclical trend to them. For example, the cost of heating oil, or corn crops—those are examples of commodities where we expect a kind of a cyclical model.

In order to account for the fact that there are trends in the underlying stock prices, a more robust kind of model was created, and Paul Samuelson was an economist who developed a model in the 1960s that included three components to it. So basically the three components are: Today's price is the first component; the second component is a component that discusses the trend—it's a drift component that talks about the trend; and then to that trend is added a random walk feature that models the volatility of the stock.

Under that kind of model we have these kinds of descriptions; for example, here's an example where we have a slight upward linear trend, together with a random walk. Here's an example of what we might expect from a stock that has a cyclical trend plus a random walk. These are models of stock prices that are based on the basic concept of trend together with this kind of random fluctuation that was described by Bachelier back in the early days of the 1900s.

And by the way, something I forgot to say, but I wanted to say, is that Einstein actually used some of Bachelier's work, which was on Brownian motion, in Einstein's description of Brownian motion. In fact, Bachelier's mathematical work on random walks was used in physics as well as in the world of finance.

We've talked about so far the issues about the stocks and modeling how stocks vary in price over time. But there's another collection of financial instruments that are called options, and option trading has become a very important part of the financial system. In fact, more money changes hands in the options market these days than in the stock market.

I wanted to spend a few minutes just describing, first of all, what an option is. An option is a contract, and the contract gives the holder of the option certain specified rights, and the rights are generally the right to buy a security or a commodity—you know, a stock or a collection of goods—at a specified price at a specified future date; or other options allow you to sell a commodity at a certain specified price on a specified date.

Now, actually options can be very complicated. They can refer to all sorts of situations, but in this lecture we're going to be talking about the simplest kinds of options, for the most part; namely, that we have a piece of paper that says the following thing on it. It says that if I hold this piece of paper I am entitled to buy one share of XYZ stock on April 30 of some future year for $100, even if at that time the XYZ stock is trading for a much higher price. And in fact that possibility—the possibility that XYZ will be worth more than $100 at that future April date—is what gives the option its value. Now, of course, if I have the option to buy XYZ at $100, and this April date comes around and the stock at that time is selling for $90, then the option to pay $100 for it is worth nothing. And that is typical of the situation for options.

Now, options can be used for two purposes. One is they can be used for speculation or, second, they can be used as hedges against a possible risk. So let me describe these two uses of options. The option used as speculation is just this: If I have the option to buy a stock at $100, I'm betting that the stock is going to go up above $100. And if, for example, on the April date on which I have the right to buy it for $100 it's actually selling for $150, then that option is worth $50 at that time because I have the right to buy it at $100, and then I can turn around and sell it for $150—so it has value to me at that time. One reason for buying an option is for speculation—I'm betting that the stock will go up in this case.

But there's another valuable, important use of options, and that is to use as hedges against risk. For example, suppose that I own a company that makes computer chips, and I use copper in the manufacture of these chips. Well, copper may be selling for a certain price today, and I have a business plan that requires that I acquire a certain amount of copper in order to do my manufacturing at a certain price. There's a danger. The danger is that in several months' time maybe the price of copper will have taken off, and then my business plan goes out the window because I have to pay much more for the materials to make the product that I'm producing.

What can I do? Well, I can buy an option as an insurance policy, basically. I can buy the option to buy copper for today's price. Suppose copper is selling for $100 per unit, well, then, I can buy an option to buy copper at $100 per unit three months from now, and that way I know for sure that I can acquire that copper at that price, and my business plan is settled. Now, of course, I pay a premium, just like any other kind of insurance premium, I have to pay a premium for the cost of that option. And in fact, that's what we'll turn to now.

How much should you pay for an option? The pricing of options, finding a rational price for options, is an idea that was developed by economists in the late 1960s and early 1970s, and really allowed the options market to be created. So let's start with a very simple option and discuss what a rational price for this option might be.

Here's an example of an option. The option is that you will pay me $1 if the stock reaches $100. In other words, I own this option. I have bought something that makes you do something else. It's a contract. Namely, you will pay me $1 if something happens in the future; namely, that, let's call it XYZ stock, reaches $100 in the future. Now, today, let's suppose it's selling for $87. My question is: How much should I pay to acquire that option to get $1 if it reaches $100? So you're in a position of trying to decide, how much do you think that I should have to pay you for that right to get $1 if it reaches $100?

Well, you might think that this is a question of probability because there is some chance that it will reach $100 and some chance that it won't. Now, you and I may have differences of opinion about what the expectation is with respect to the future of this stock. For example, I might think that there is a good chance that the stock will reach $100, and so to me, having the option to get $1 if it reaches $100 is something good. And I could give it as an expected value analysis. For example, if I thought that the probability of its reaching $100 was, let's say, oh, 90%, then I can do an expected value computation, and say, "Well, okay, I'm going to get $1 90% of the time; therefore, 90 cents sounds like a good price for that option to me." That's what it would be worth for my concept, my guess, of the future.

But a different person might have a different guess of the future. For example, you might feel that the stock has a 50% chance of never reaching $100. So if you do your expected value analysis, you would say, "Well, I have a 50% chance of having to pay $1, but I have a 50% chance of never having to pay the dollar. Therefore my expected value of that option, the price for it, would be 50 cents."

Now, those differences of opinion about the value of the option to be given a $1 if the stock reaches $100, that difference makes it why a deal is to be made. But there's a rational way to set a price for this option, and that is to think about the question: Can we get rid of the probability? Can we get rid of the risk in some way? And the answer is yes. We can do something today that will completely get rid of the risk of what it would cost to pay $1 if the stock reaches $100; namely this: suppose that you were contemplating selling me this option, so that I would get $1 if it reaches $100, and you were sitting there and saying, "Oh, I don't really want to worry about that. I don't want to worry having to pay this $1 if something happens, so what can I do today to make sure that I'm going to have the resources to pay off that $1 if the stock reaches $100?"

The answer: very simple. If you buy 1/100 of a share of the stock today, then you own 1/100 of a share of that stock, so if the stock goes up to $100, 1/100 of that stock is worth $1, you see? So you can do something today that gets rid of the risk. It gets rid of the risk. So that is a basic concept about options pricing. What can you do today that does not involve risk that duplicates the risk that you entail by owning the option? So the rational price for the option is going to be 87 cents—1/100 of the price of one share of stock—and that's a rational price for this option.

We're now going to look at an example that's a little bit more interesting in that there are more variables involved, but still a great simplification relative to actual reality. Here's the example we're going to be discussing. Suppose that we're talking about an option associated with a stock that today is selling for $100, and we're talking about the option of buying the stock one month from today for $100, the current price of the stock. You understand what the option is, the option would be that I have the right to buy one share of stock for $100 one month from now. Now, I don't know whether the stock is going to go up or going to go down, but we're going to make a very simplifying assumption, and the simplifying assumption is that there are only two choices for what will happen to the stock in one month's time. Either it will be $110 or it will be worth $95 one month from today.

Now, by the way, this is, of course, a great simplification, because in fact we don't know. It could be any price, both between $110 and $95, or it could be greater than $110 or less than $95. But this is the kind of mathematical simplification that allows us to develop tools that then can be applied in the more general cases. In fact, this concept of just looking at a finite collection of possible future values at discrete moments of time is called the Cox-Ross-Rubenstein tree, and this was a kind of a model for stock prices that allow us to develop a rational pricing system for options.

So we're discussing the question of how much should an option be worth that gives me the right to buy a share of stock at $100 one month from today. And the simplifying assumption again is that the price will be either exactly $110 or exactly $95, and those are the only two choices. Well, here's how we think about it. What we're trying to do is to replicate the risk of the option. In other words, we're trying to buy some collection of shares of stock, just like in the last example we said, "If we own 1/100 of a share of stock, then we don't have to worry about the risk." We're going to do the same thing here. We're going to say, "I'm going to buy a certain number of shares of that stock and put it in my portfolio, and then I'm also going to have a certain amount of cash in the portfolio," and by the way, the amount of cash is going to be negative in this case; so that is, you'll think about borrowing money. But you'll have a particular portfolio whose value is going to be exactly equal to the value of the option in one month's time. That's the concept. In other words, can we own something today that gets rid of the risk? So this is an example where we're trying to actually quantify the value of risk itself—and that's an interesting perspective.

Here we go: Our goal then is to buy x shares of stock in the portfolio, and have d amount of cash in the portfolio, so that it replicates the value of the option in one month's time. Let's think about it. The price of the stock we're assuming is either going to be $110 in one month's time or it's going to be $95 in one month's time. If the price goes to $110, then how much is the option worth to buy the stock for $100? Well, the option is worth $10, because you can buy the stock for $100 and immediately sell it for $110. So the value of the option in one month's time may be $10—if the stock goes to $110. On the other hand, if the stock price descends to $95, then the option will be worth nothing.

What we can do is write down two equations that capture the reality of the costs of the option. In other words, if we buy x number of shares of stock, and the stock were selling for $110 in one month's time, and we have d dollars, we want to replicate the value of the option, so we want that number to equal $10. And on the other hand, if we own that same x number of shares of stock, and the price of the stock in one month's time is $95, and we have our d dollars in cash, we want that value to also equal the value of the option in one month's time; namely, 0, because the option would be worthless if the stock is only selling for $95.

Consequently, we have two equations and two unknowns, and we can just do a little bit of algebra and actually compute what the values are. So here we take these two equations and two unknowns, we do a little algebra, and we find that the solution is that we should buy 2/3 of a share of stock and we should borrow $63.33. In other words, by the way, we should have on hand –$63.33, which means that we're borrowing that amount of money. So if we have that portfolio, that portfolio that consists of 2/3 of a share of our stock and we owe $63.33, then that portfolio has the same value as the option one month from now.

Now let me just say one other simplifying assumption that we're making here, and that is that there is no cost for money. We're not assuming that you're earning interest or not earning interest. That's just a simplifying assumption that would have to be factored in, in reality. But here we go. Then what that says is that the cost of getting rid of the risk associated with the option is the cost of 2/3 of a share of stock minus $63.33. Well, 2/3 of a share of stock would cost $66.67, or $66.666 (2/3 of $100). On the other hand, borrowing $63.33 gives a total cost of the option of $3.33.

In other words, this is a way of analyzing a rational price of risk itself, the risk that's entailed by owning this option, of having to perform the option. In other words, having the option to sell the stock at $100 even if it goes up to $110, that we have made a model which allows us to say that for $3.33 we could replicate the risk by something that we could just own, and now there's no risk to it. If we own this 2/3 share of stock, regardless of whether this stock goes up to $110 or goes down to $95, the value of our portfolio is exactly equal to the value of that option. And the cost of that was $3.33.

Well, of course, the model that we've described here is a great simplification, and it had to be developed and generalized to the much more robust example of a complicated scenario of real life. But the mathematicians who created this kind of a model generalized it to give rational pricings to options, and this kind of analysis leads to the famous Black–Scholes model of options pricing. After the Black–Scholes model came into existence, well that's what allowed options to be considered as investments rather than as gambling, and that led to the options trading houses, like the Chicago Board Options Exchange, which came into existence in 1973—really rather recently.

But there's a story associated with this options trading, and that's the story of the Long-Term Capital Management Company (LTCM); this was a hedge fund that was run out of Greenwich, Connecticut, and was founded in 1994. The man in charge of this hedge fund was John Meriwether, who was a legendary head of bond trading at Salomon Brothers in the 1980s. In addition to the many mathematicians that he brought from Salomon to LTCM, he hired Myron Scholes and Robert Merton to be on the Board of Directors, and both of them, by the way, went on to win the Nobel Prize in Economics in 1997 for their work on options pricing.

Well LTCM used complicated mathematical strategies to trade bond products. It had sophisticated models that were used to describe how certain future bond prices would converge or diverge, and LTCM invested huge sums of money into these positions. LTCM had a really rather large start up of over a billion dollars, and it quickly grew to 4.7 billion dollars in equity in just three years. It was a money-making juggernaut. So to take full advantage of the bond mispricings found by their models, LTCM borrowed, they borrowed immense amounts of money. In other words, they leveraged their money to fund their trades, and the company just spun out of control. They borrowed over 125 billion dollars, and in fact their balance sheet had derivative positions—derivatives or options; derivative positions over 1.25 trillion dollars.

Let me explain their strategy. Their strategy was to bet on the relative difference in the prices of two types of bonds. So a standard risk arbitrage trade that LTCM used was to bet on the convergence of two prices in the market. Say we're looking at foreign bonds versus U.S. Treasury bonds—the prices of those two groups of bonds—and suppose our mathematical models—and they had lots of statistical backing to them—suppose these models tell us that the spread—that is, the price difference—between these two collections of bonds should be, say, 50 cents, but in fact the spread right now is $1.

Well, according to our modeling, then, the price should close. Now we don't know whether the prices are going to go up or go down, because the prices themselves depend on decisions made by politicians and others that are very hard to predict; but nevertheless, we can believe that the prices will come closer together, whether or not the underlying price of the bonds goes up or down.

What we can do is we can invest in options in such a way that as long as the prices come closer together, we make money; this was the thinking of LTCM. And the further the spread widens, the more trouble that the traders would find themselves in, and that's what happened at LTCM. Things began to unravel for them in about 1998. At that time what happened was Russia defaulted on its ruble debt, which was viewed as very improbable event, and as a result investors around the globe sold European bonds in favor of buying more stable U.S. bonds.

This caused various bond prices to diverge, whereas LTCM had bet hundreds of billions of dollars that they would converge. Because LTCM had borrowed so much money, they couldn't afford to wait for the bond prices to eventually converge, and so it became a problem of solvency. What happened is they didn't remain solvent long enough to ride out the spread of the bond prices, and wait until they reconverged as they had predicted.

In the end, in fact, the Federal Reserve Bank of New York had to intervene, and it arranged a bailout of several billion dollars by 14 major investment banks, and in all, LTCM managed to lose $4.6 billion dollars in three months. Well, in fact, the bond prices did ultimately converge in 1999; but of course it's an old story, that you have to have the capital enough to ride out the losses until they become wins.

Well, this was a discussion of gambling and randomness and probability in the world of finance. In the next lecture we're going to turn our attention to finding probability in unexpected places. I'll see you then.

Lecture Eight
Probability Where We Don't Expect It

Scope:

Sometimes, probability is centrally involved in solving problems that seem to have no random or probabilistic component to them at all. In mathematics, an example occurs in some methods of determining whether a number is prime or not. Any number is either prime or it's not—there is no randomness involved—yet probabilistic methods can essentially determine whether a number is prime even when the number is far too large for any computer to factor. Randomness and probability are involved in psychology when talking about conditioned behavior. Pigeons rewarded randomly rather than on any fixed pattern will retain their training longest. Strategic decision-making, or game theory, often finds that optimal strategies involve taking one action or another with a certain probability rather than finding one best move. Optimal business strategies or sports strategies often are probabilistic rather than deterministic. Probability pops up in many unlikely places.

Outline

I. In this lecture, we will talk about finding probability in unexpected places. We start with the world of math.

II. Probability can be used to determine to any desired degree of certainty the primality of a natural number with hundreds of digits.

   A. Whether a positive whole number is prime (that is, whether the number is not the product of natural numbers smaller than itself) is clearly not a question with any random or undeterministic feature, yet a method of determining whether it is prime uses randomness and probability.

   B. One way to see if a number is prime is to try to divide into it all smaller numbers. Here is an example of this method of determining that the number 91 is not prime.

Divide 91 by	Get remainder of
2	1

3	1
4	3
5	1
6	1
7	0

1. When we arrive at 7, we see that 91 divided by 7 is 13, with no remainder; thus, 91 is not prime.
2. This strategy, however, would be impossible to use for longer numbers, even with today's computers.

C. The method is effective even when it might be impossible to determine whether or not the number is prime in any known deterministic way.

D. Another strategy for determining if a number is prime uses *Fermat's little theorem*: Start with a number that is prime, take any number less than that number and raise it to the power of 1 less than the prime, then divide by the prime; you get a remainder of 1. This remainder formula is written $n^{p-1} \equiv 1 \bmod p$.

1. For example, if you start with the prime number 5, then you take any number less than 5 (for example, 2) and raise it to the fourth power (5 − 1), you get 16, and $\frac{16}{5} = 3$, *with a remainder of 1.*

2. Likewise, if you start with the prime number 5, then you take 3 (instead of 2) and raise it to the fourth power (5 − 1), you get 81, and $\frac{81}{5} = 16$, *with a remainder of 1.* If you start with the prime number 5, then you take 4 (instead of 2) and raise it to the fourth power, you get 256, and $\frac{256}{5} = 51$, *with a remainder of 1.*

3. Let's take a different prime, 7. If we choose 2 as the smaller number, then we find 2^6 is 64, and $\frac{64}{7} = 9$, *with a remainder of 1.* No matter what smaller number we choose, we always have *a remainder of 1.*

III. This theorem then gives us a way to see if a number is *not* prime.

    A. For example, we can prove 9 is not prime:

$$2^8 = 256$$

$$\frac{256}{9} = 28, \text{ with a remainder of } 4$$

        1. Because the remainder is not 1, 9 is *not* prime.

        2. In addition, there is a computational simplification using just remainders that speeds up the calculation.

    B. We can also use this theorem to test if a huge number is not prime. The question must be asked, though: Even if we use the number 2, how do we raise it to the required power and find the remainder after dividing?

        1. For large values of p, $2^{p-1} \bmod p$ can be cleverly computed by simplifying:

$$2 \times 2 = 2^2$$
$$2^2 \times 2^2 = 2^4$$
$$2^4 \times 2^4 = 2^8$$
$$2^8 \times 2^8 = 2^{16}$$
$$2^{16} \times 2^{16} = 2^{32}, \text{ etc.}$$

        2. If p has 300 digits, it takes only on the order of 1,000 such doublings to calculate $2^{p-1} \bmod p$.

    C. This is a probabilistic test, however, because some numbers fool it.

        1. For example, 341 is a product of 11×31, yet 2^{340} divided by 341 does give a remainder of 1.

        2. However, for a randomly chosen 13-digit number, there is a 99.9999985% chance that a number that satisfies this test is prime. Of the 308,457,624,821 thirteen-digit primes, only 132,640 will fool this test!

IV. Probability arises in game theory.

    A. Game theory is the mathematical model of strategic decision-making. It is used in economics, business, games, sports, war, and other areas where strategic decisions must be made.

    B. Game theory uses the concept of a payoff matrix, which describes the payoffs for each player for each combination of options that the players could choose.

V. We will study game theory as it applies to football.

    A. In football, on the third down with many yards to go for a first down, the usual options are a pass play or a run play.

    B. The defending team, then, can defend against the pass or defend against the run.

    C. Below is a possible payoff matrix. Each number represents expected yards gained by the offense. The defensive payoffs are understood to be the negative of the numbers:

Defense Options → Offense Options ↓	Defend against Pass	Defend against Run
Pass	5	7
Run	6	1

    D. If the offense always passes, the defense will learn to always defend against the pass. That combination gives an expected value of 5 yards for the offense. But if the offense always runs, the defense will learn to always defend against the run. That combination gives 1 for the offense.

        1. Game theory confirms that once in a while, at random, making the unobvious play is the best long-run strategy.

        2. According to our calculations, the expected number of yards gained if the offense passes with probability p and the defense defends against the pass is $p \times 5 + (1 - p) \times 6$.

        3. The expected number of yards gained if the offense passes with probability p and the defense defends against the run is:
$$p \times 7 + (1 - p) \times 1.$$

        4. Our probability of passing is a max/min strategy:
$$p \times 5 + (1 - p) \times 6 = p \times 7 + (1 - p) \times 1.$$

        5. Our conclusion is that the offense should pass 71% of the time (randomly).

        6. That combination gives an expected value of 5.3 for the offense, which is a higher value than either of the two pure strategies.

7. Likewise, using the payoff matrix figures again, we find that the defense should defend against the pass 86% of the time (randomly).

8. This is called a *Nash equilibrium*, that is, a strategy whereby no player can get an advantage by unilaterally changing strategy. It was named for John Nash, who won the Nobel Prize for his work on game theory.

VI. Let's turn now to risk management in business, studying how a large NASA project estimates its budget.

    A. The project lists all the risks that might incur a cost, with an estimate of both the possible cost and the probability of occurrence. The expected value is the probability of occurrence times the cost.

    B. As risks are retired or reevaluated or as new risks are added to the list, the expected value is recomputed.

    C. In this way, the project can estimate how much money it should keep in reserve.

VII. Psychologists have learned that randomness can play a valuable role in reinforcing a desired behavior.

    A. Giving rewards is an ingredient in training an animal, for instance, a pigeon, to behave in a desired way.

    B. The question is, how frequently should you reward the instances of the desired behavior in order to have the conditioning last the longest?

        1. If you give a reward for a certain behavior (pecking) every time, at first the pigeon learns but quits rather soon when the reward ceases to appear.

        2. The best strategy is to randomly reinforce the behavior. Changing the frequency of rewards in an unpredictable, random way leads to behaviors that are retained for long periods even in the absence of rewards.

    C. Applied to humans, this observation may help explain the compulsiveness of some gamblers.

Readings:

Edward B. Burger and Michael Starbird, *The Heart of Mathematics: An invitation to effective thinking*, 2nd ed.

Oskar Morgenstern and John von Neumann, *Theory of Games and Economic Behavior* (commemorative edition).

Questions to Consider:

1. Here is a payoff matrix in which each player is a driver who can choose to drive on the right or on the left of the street.

	Drive on Right	**Drive on Left**
Drive on Right	70,70	−100,−100
Drive on Left	−100,−100	70,70

The value (by some measure) is 70 for each player if the two agree to drive on the same side of the street, thus avoiding a crash when going in opposite directions. The value is −100 if they choose different sides and, thus, crash. What are the Nash equilibriums for this payoff matrix?

2. Can you think of an example in your own life where random reinforcement has had a lasting impact on your behavior or attitudes?

Lecture Eight—Transcript
Probability Where We Don't Expect It

Welcome back. In the previous lectures what we've heard about are instances in which probability and randomness seem pretty clearly to apply. We've talked about gambling where, obviously, randomness and probability come into play. In the case of physics, we can easily imagine that all these molecules moving around randomly have a random component and probabilistic component to them. We talked about the genetic drift and randomness was involved in genetic inheritance; so these are clearly examples where randomness plays a role.

In the last lecture we talked about the world of finance, where stock markets, of course, will have a random component to them, but all of those are places where we sort of expect probability to play a role. Well, in today's lecture we're going to talk about finding probability in unexpected places—places where you wouldn't expect probability to play a role at all, and we're going to start with an instance that in my world, at least, I think of as maybe the place where you would really not expect any probability to play a role; namely, suppose that I write down a specific number, just the digits of a number, and I ask the question: Is that number prime or not?

Now surely no probability could be involved with such a question—either it's prime or it's not prime. There's no issue here. But it turns out that there are strategies for determining the primality of huge numbers and deciding whether they are prime that involve probability.

So let's begin, though, by reminding ourselves what a prime is and how we would naturally try to find out whether a number is prime. So let's take a specific example. First of all, let me remind you what a prime is. A prime is a positive whole number, a natural number that cannot be written as the product of smaller numbers than itself. So it's some number, 2 or higher, that is not written as a product of smaller numbers. So, for example, the number 5 can't be written as smaller numbers, so it's prime; seven cannot be written as smaller numbers, so it's prime; six is not prime because it's (2×3). So that's what a prime is.

Now, how would we determine whether something is prime or not? Well suppose we just take a number, and I'll take the number 91, and let's try to decide whether or not the number 91 is or is not a prime. Well, since the definition of being prime has to do with its divisibility by smaller numbers, let's just try to divide it and see whether or not we get a remainder; and by the way, the reason that I choose 91 as an example is because some people view 91 to be the smallest number that looks prime but isn't. I don't know if you think that way; it looks sort of prime to me, but it's not. Okay, so here's how you would go ahead to find out whether 91 is prime. You try dividing. You divide by 2, you get a remainder of 1; divide by 3, you get a remainder of 1; divide by 4, 4 goes into 88 evenly, so it has a remainder of 3; divide by 5, you get a remainder of 1; divide by 6, get a remainder of 1; then you get to 7 and it goes in evenly, and 91 = 7 × 13.

That's a wonderful strategy for determining that 91 is not prime. Now I want you to try it with this number. Is this number prime? Well, this number has 300 digits to it—300 digits. How in the world would you determine whether that's prime? You might consider trying the same method; that is, just start dividing by 2, see if there's a remainder; 3, see if there's a remainder, and so on. The trouble is that if you tried that strategy, and even if you could attempt divisions at the rate of, say, oh, a trillion divisions per second, if you started at the time of the Big Bang, 13.6 billion years ago, and you did a trillion divisions per second, you would not even be close to doing all the divisions that you would need to determine whether this number is prime, not even close. Unless you have much longer than the history of the universe, that method will not work.

We have to think of something else, and it turns out that one of the ways to look at it is to use a theorem that was proved by Fermat. Now you remember Fermat. I mentioned Fermat as being one of the originators of the concept of probability, but this is not a probabilistic theorem—and you also remember Fermat from the famous Fermat's last theorem, which was proved recently, after 350 years. But this is a theorem that Fermat proved, called Fermat's little theorem. And it's the following statement. It says that if you take any number that actually is prime, and you take any number less than it—whole number, we're just talking about integers, so you take any integer less than that prime, a positive integer—and you raise it to the power of 1 less than that prime, multiply it by itself that many times, divide by the prime, you get a remainder of 1.

Now, I said that sort of fast, so we'll just do it some examples so you'll see what it means. This says that if, for example, you take the number 5; 5 is a prime. If you take any number less than 5, like, let's say 2, you raise it to the fourth power and divide by 5, after dividing, you'll get a remainder of 1. In this case, 2^4 is 16, divided by 5 goes 3 times, with a remainder of 1. The same thing if you take 3 and raise it to the fourth power—you get $(81 \div 5)$, you get 16 with a remainder of 1. Take the number 4. You take 4 to the fourth power, you get 256; you divide by 5, it goes in 51 times with a remainder of 1.

These are just examples of Fermat's little theorem. Let's take the prime 7 $(2^6 \div 7)$, you divide in and you get a remainder of 1; $(3^6 \div 7)$, divide in, you get remainder of 1; and so on. It's true for every single prime; I could prove it, if we took half an hour we could prove that theorem. It's not a hard theorem to prove, but it is an interesting theorem, and it's true for all primes. Well, the truth of Fermat's little theorem gives us a method for determining whether a number is not prime, and the way it can determine whether a number is not prime is this, that you just take any number...

Let's be specific and prove that the number 9 is not prime; 9 is (3×3), but let's suppose we didn't notice that, and instead we wanted to use Fermat's little theorem to prove that 9 is not prime. What we would do is we would take $(2^8 \div 9)$ and see what the remainder was. Now, Fermat's theorem would say that if 9 were prime—which it's not, of course, but if it were prime—then $(2^8 \div 9)$ would have a remainder of 1. Well, when we actually do it $(2^8 \div 9)$ is $(256 \div 9)$, has a remainder of 4, not 1; and since it's not 1, we know that 9 could not be prime.

And by the way there's a technical point I want to just say quickly here, that we're trying to take 2 to the eighth power. Well, 2^8 is ($2^4 \times 2^4$), which is (16×16). Now, it turns out that if you wanted to, you could just use the remainders of these—instead of multiplying 16 together, you could just take the remainder upon division by 9, 7. ($7 \times 7 \div 9$) will have the same remainder as $(16 \times 16 \div 9)$. So it's a simplification so you don't have to multiply the bigger number 16. You could just multiply the remainders together and get the same remainder; but that's a technical point.

Okay. So here is the point: If we had this huge 300-digit number and we're trying to figure out whether or not it's prime, let's just take the number 2, raise it to the power of one less than this huge number, divide by the huge number, and see if we get a remainder of 1. If we get any remainder other than 1, we know it's not prime.

Now you might say, "Well, how in the world are you going to take 2 to the power of a 300-digit number? You'd be multiplying..." If I said that it took more than the history of the universe time to divide, it would take more than that history of time to multiply. Well, the answer is that you do it cleverly. You just take (2×2), then that's 2^2, and then ($2^2 \times 2^2$ is 2^4; $2^4 \times 2^4$ is 2^8; $2^8 \times 2^8$) is 2^{16}, so you quickly have those exponents growing. They're doubling each time there's a multiplication. So in that fashion, you can actually raise a number even to this 300-digit number that you want with only about 1,000 such doublings, and so that's well within the scope of just a little computer, and so you can quickly determine what that remainder is.

The answer is that this huge number is, in fact, a prime; but it's a probabilistic test, because there are some numbers that fool this test. For example, the number 341—the number 341 is (11×31), and yet if you take the number 2 and you raise it to the 340th power and divide by 341, you do get a remainder of 1, so it would fool this test. But the test being just to take the number 2, raising it to one less than this number power, dividing the number and finding the remainder. Well, it turns out that although the test failed for 341, it turns out to be a very good test in the sense that the probability of being fooled by the test gets more and more remote as you get bigger numbers, and that the probability is pretty darn good.

For example, among 13-digit numbers: If you randomly choose a 13-digit number and you do the test of raising the number 2 to the power of one less than the number, dividing by the number, and seeing whether you get a remainder of 1, you can get a remainder of 1 even for some non-primes, but your probability of doing so is extremely remote. The probability of such a number actually being a prime is 99.9999985. That would be your chance of actually having it be a prime, if you just did that one test. It turns out there are 308,457,624,821 13-digit primes, and of those, only 132,640 will fool this test of 2 raised to that power. Now how do people know these numbers? Well, that's what number theorists are for; they figure out these things.

Okay. So this is an example where probability comes into play to figure out the primality of a number. That seems like a strange place for probability to play a role.

Well, let's find another situation. We're going to just do several different situations where probability comes into play in unexpected ways. This next one involves game theory. Game theory is a mathematical model of situations like games, where the players use different strategies to maximize their return. And so what we're going to do is to see that probability arises, I think rather unexpectedly, in game theory. By the way, game theory is used in economics and in business; it certainly is in games, of course; but it's also used in sports, in war, modeling war situations, lots of areas; wherever strategic decisions are made. It uses the concept of a payoff matrix where you look at the options of one player and you correspond them to the options of the other player, and see what the payoff is under those circumstances.

Now, we're going to be investigating an example of such a payoff matrix in the world of football. So here we are; we're going to play a football game, and we're imagining that we have the ball and it's third down with long yards to go, and we're the offense. We're saying to ourselves, "I have a choice. I could either do a pass play or a run play." Now, by the way, you don't have to know much about football for this. We're really talking about the math. Okay, so you can either pass or run. Now the defense can either defend against the pass or defend against the run.

We could write down a matrix, then, that captured the expected number of yards that the offense would gain on average given the four different possible alignments of pass/run on the offense, defend against pass, defend against run on the defense. So this chart says that if the offense passes and the defense defends against the pass, on average the offense will make 5 yards. On the other hand, if the offense passes and the defense is aligned against the run, the offense will make 7 yards.

Here this square corresponds to the offense running while the defense is aligned against the pass. The offense in that case makes 6 yards; but if the offense runs and the defense defends against the run, the offense makes only 1 yard on average. So this is a payoff matrix for this decision about whether to pass or run, or defend against the pass or run.

Now, in trying to decide how to use this matrix to make a decision about whether to pass or run, let's imagine that we're the offense. We say to ourselves, "Well, if I always pass, and the defense always defends against the pass, then on average I'll make 5 yards. But if the defense is defending against the pass, what I'll do is I'll occasionally run. I'll fool them, you see, because they're defending against the pass. So then my expectation would be to make 6 yards instead of 5, if I run when they're aligned against the pass." But, of course, knowing that, the defense then won't always defend just against the pass, because if it always said, "I'll defend against the pass," then the offense could choose to run, you see?

It's this kind of thinking process that leads to the idea that instead of choosing one strategy, of always passing or always running, the best strategy for the offense is to choose to pass with a certain probability. In other words, sometimes you're going to pass; sometimes you're going to run; and you're going to randomly choose which to do according to some weighted probability.

The question to ask is the following. Suppose we ask ourselves the question: What are the expected number of yards that the offense would be expected to gain if it passes with probability P, and the defense is aligned against the pass? Suppose the defense is aligned against the pass, and for every probability P we can ask ourselves the question: What is the expected number of yards that the offense will gain?"

Well, looking at this chart we see that if with probability p we pass, that means with probability $1 - P$ we will run. So the expected value for a particular probability p—the expected number of yards gained—is P (the probability of a pass) × 5 (the number of yards we expect to gain if we pass, and if the defense is defending against the pass) + $(1 - P)$ (which is the probability of running) × 6 (which is the expected number of yards that we will gain if the defense is aligned against the pass). This gives an equation that tells us the expected number of yards we will gain in the offense for every choice of probability P between 0 and 1 if the defense is aligned against the pass.

Now, in the same manner, if the defense is aligned against the run, we get a similar equation, but here, every time we pass and they are defending against the run, the chart tells us that if we pass and they're defending against the run, we will make 7 yards on average, that means that the expected number of yards that we will make if the defense is defending against the run is $[P \times 7 + (1 - P$—the probability of running$) \times 1]$ because if they're defending against the run, we'll only make 1 yard on average.

Now, let's look here. We're the offense and we're thinking about what to do. So this is the offense's perspective of the problem; namely, for each probability P, we get a number of the expected yards we would gain if the defense is defending against the run—that's this blue line—or if the defense is defending against the pass—that's this red line. So I claim that the percentage of the probability with which the offense should pass is exactly at this point right here. And the reason that this is the percentage—you see the percentage is between 0 and 1, it's the probability of passing—that's the decision the offense is making: With what probability should I pass?

The reason that the offense should choose this probability of passing is the following. Suppose instead the offense chose to pass, say 50% of the time—down here; then if the defense always defended against the run—namely on this blue line; the expected number of yards would be down here. It would be a lower value than the expected number of yards here. In other words, if the offense only passes 50% of the time, the defense could do something—namely, always defend against the run—that would lower the expected yards that the offense is expected to make.

Likewise, suppose the offense chose to, say, pass every time. If it chose to pass every time—pass with probability 1—well, then, the defense can always defend against the pass, and the expected number of yards would, once again, be lower. You see this line is going downward, so the expected number of yards gained for the offense would be less. It's only at the place where the defense against the pass and the defense against the run cross, that's the place where we should, as the offense, choose as our probability of passing, because that maximizes the minimum gain that we're expected to make. It's a max/min strategy.

Well, first of all, how do we actually compute that value? Well, it's easy. We say it's the place where the expected value of the number of yards gained if the defense is defending against the pass is equal to the expected gain that the offense makes if the defense is defending against the run. Those are the two equations that we previously computed as our expected value for any given probability P of yards gained. If we then just do the algebra, we set those two things equal, and then solve for P, we see that P is exactly 5/7, which is 0.714 and so on. So that means that the offense's best strategy is to pass with probability 71%.

So, isn't that interesting? It's a place where probability comes in as the method of finding out how to maximize the expected number of yards gained by the offense—and, by the way, I should say that if the offense does pass with a frequency, with a probability of 71%, then it maximizes the expected gain, and it's exactly the same whether the defense is aligned against the run or aligned against the pass. It's 5.3 yards on average, is the amount that you would gain. And if you tried any other strategy, if the offense did any other percent strategy, then the defense could do something that would lower that average gain; this is a max/min strategy.

Now, what happens? Well the defense does exactly the same kind of computation. It computes its expected value for picking a probability of defending against the pass, so it's making the same kind of conceptual decision, and it does the same kind of computation, and it discovers that it should align itself against the pass with a probability of 86%.

Interestingly, both the offense and the defense have a percent that they should randomly choose—on the offense's side to pass, and on the defense's side to defend against the pass—with the probabilities respectively of 71% for the offense—is the percentage of time the offense should pass rather than run—and the defense should randomly defend against the pass 86% of the time.

It's interesting because what this means is that, first of all, this is a mixed strategy in the sense, mixed, meaning that it's a probabilistic strategy. It's not just always pass or always run, or always defend against the pass, or always defend against the run—it's a mixed strategy, a probabilistic strategy. And so it has the property that both the offense and the defense might, from time to time, be misaligned, and that's what sort of makes it fun.

In ordinary jargon about actually watching football games, you can say things like this, "Oh, the offense ran in order to keep them honest," and that's approximately true. And the defense has to do the same thing—occasionally defend against the run even though there's a high chance that the people will be passing. And then, of course, it gives commentators a great opportunity to say how dumb the coaches were to make these decisions if they happen to be misaligned.

But these probabilities are an example of what's called a Nash equilibrium. A Nash equilibrium is a strategy with the property that no player can get an advantage by unilaterally changing his or her strategy. In other words, the offense can't change the strategy and gain an advantage, and neither can the defense. Nash was a mathematician who won the Nobel Prize in economics for his work on game theory, and this existence of such a Nash equilibrium such as we've just seen in this example is the kind of thing that he won his Nobel Prize for, and he became famous for this work because of the book, *A Beautiful Mind*, and the movie, *A Beautiful Mind*.

Anyway, I thought the idea that in making strategies of games that we use randomness instead of just a direct strategy, I thought that was an interesting, unexpected place for randomness and probability to occur.

Well, let's turn our attention from game theory to risk management in business and in managing large projects. It turns out that NASA makes these very big projects that send space exploration, going to Mars and other places, and these are very big projects, and they involve all sorts of aspects where there's uncertainty. People don't know what's going to come up because they're doing things that have never been done before. And so you have a large number of features of the project that have risks to them, and unknowns.

Well, in deciding how to describe their budget, this is what NASA actually does. What it does is that the project lists all the risks that it can think of that would incur some sort of cost, and then it estimates two things about those risks. It estimates the possible cost of overcoming that problem, and then it also estimates the probability of the occurrence of that risk. Then you have the opportunity to take an expected value, you see, because you have probability of a risk occurring times the cost that will be incurred if that happens.

In other words, if, for example, a particular risk will cost an extra $100,000,000, but it has only a 30% probability of happening, then that would be evaluated as a $30,000,000 risk as the expected value—it may cost $100,000,000 or it may cost nothing, but on average it's that $30,000,000. By breaking up the project into small pieces, you can have a whole collection of risks with their expected costs. Then as the project continues some of the risks are retired, meaning that you've gone beyond the point where that risk happened or didn't happen, so then you can retire that risk from the list and reevaluate your expected extra money that you need in reserve. So this is a way for a big project to manage how much money it should have in reserves for the project.

That's an interesting place, I think, where probability plays a role in management, because the unexpected does happen, and so how do we deal with the fact that unexpected things do happen? It's not good enough just to say, "Oh, stick to your budget," when you're doing something where you don't know what's going to happen.

Well, another example where randomness comes in is in the area of psychology. You all know about Pavlov's dogs and the operant conditioning. Well, one question is: How should you condition an animal so that their learning lasts the longest? In other words, that they will learn something and will not lose that learning over the longest period of time. What's the strategy of rewards?

Let's imagine, say, a pigeon that's being trained to peck at a little dot, and when it pecks at the dot a particle of food comes out, a piece of food comes out, and the pigeon eats it and is happy. It likes the food. So the question is: With what frequency are you best off giving the pigeon the reward in order to have the pigeon learn that process and keep knowing it for the longest; in other words, keep pecking at that dot?

Well you might think that giving a reward every single time would be the way to do it, because then the pigeon is reinforced most often. But the problem with that is then when you cut off the rewards, the pigeon has the expectation of getting a reward every time, and when it quits getting rewards, then its concept that I got a reward every time I did this fails, and so it quits—quits rather soon. Now, you might think well, okay, if you did it every third time, or every fifth time, or every seventh time, that would work.

It turns out that the best strategy is to randomly reinforce the pecking—to give reinforcements randomly—and in that way the pigeon doesn't know whether, when it keeps pecking, it might just be in one of these long sequences where there don't happen to be any pellets coming, you see, just because randomness has that feature, and the pigeon learns that. Well, it turned out that in actual experiments, they did some random reinforcing of pecking at a dot for a pigeon, and they did this for just one minute—for one minute. And they randomly gave him—I guess they must peck pretty fast to only do this for one minute—but this poor pigeon pecked away and got randomly reinforced, and it took 3.5 hours before the pigeon gradually stopped pecking at that dot, because it didn't know when is the next reward going to come.

It turns out, by the way, I think that this is part of the explanation for the very recalcitrant addiction to gambling, because, you know, gamblers are randomly being rewarded, and their random rewards are sometimes big. They're getting these big rewards at random intervals, and so they're induced to continue on and try yet once again to try to gamble. I think that may account for examples of gambling addiction, and other human examples of things where we get inclined to do things over and over again.

In this lecture we have seen probability in unexpected places, and then in the next lecture we're going to look at some of the really most famous examples of probability: the birthday problem and the Monty Hall game show, *Let's Make a Deal*[R]. I'll look forward to seeing you then. Bye for now.

Lecture Nine
Probability Surprises

Scope:

Probability is full of surprises. The birthday problem is a famously counterintuitive result; namely, if 50 random people are in a room, there is a 97% chance that at least two of them have the same birthday. The analysis of that probability illustrates strategies of combining probabilities. This counterintuitive result can be confirmed by looking at various groups of people, such as presidents of the United States or Oscar winners, and finding birthday coincidences as predicted. The Monty Hall problem is equally baffling to most of us. It is a challenge for all of us to hone our sense of probability so that our intuition more closely accords with reality. Tricky probability problems arise in issues from birthdays to game shows to tennis to choosing socks from a drawer!

Outline

I. Let's start this lecture with the famous birthday problem, mandatory for any probability course.

    A. If 50 random people are in a room, what is the probability that two of them will have the same birthday? In fact, the surprising answer is that there is a 97% chance that two of them will have the same birthday.

    B. It is easier to compute the probability that all 50 birthdays are different.

        1. To compute the probability that all the people have different birthdays, you would multiply as follows:
$$\frac{365}{366} \times \frac{364}{366} \times \frac{363}{366} \cdots \frac{319}{366} \times \frac{318}{366} \times \frac{317}{366} = 0.03$$

        2. The product of all the fractions is about 0.03. Thus, the probability that no two people have the same birthday is only about 3%.

    C. Hence, the chance that at least two people *do* have the same birthday is about 97%.

    D. The surprisingly high probability for birthday coincidences can be tested in reality by looking at some collections of about 50 people.

1. Of the first 42 different presidents, one pair has the same birthday: Polk and Harding: November 2 (1795, 1865). One pair and one triple of presidents have the same death day: Fillmore and Taft: March 8 (1874, 1930); J. Adams, Jefferson, Monroe: July 4 (1826, 1826, 1831).

2. Of the first 46 vice presidents, three share a birthday: Hannibal Hamlin: August 27, 1809; Charles G. Dawes: August 27, 1865; and Lyndon B. Johnson: August 27, 1908.

3. Of the Oscar winners for best actor, two have the same birthday: Ben Kingsley, who won in 1983: December 31, 1943; and Anthony Hopkins, who won in 1992: December 31, 1937. Two winners have the same death day: Humphrey Bogart, who won in 1952: January 14, 1957; and Peter Finch, who won posthumously in 1977: January 14, 1977.

E. If you have 90 random people in a room, chances are .999993 that at least two will have the same birthday. And if you have only 23 people in the room, the chances are even that at least two will have the same birthday.

II. Another famously non-intuitive problem is the Monty Hall problem from the TV show *Let's Make a Deal*[R].

A. Here is how it works:

1. A contestant in a game show gets to pick one of three doors and keep whatever prize is behind the door. One of the doors has a desirable prize; the two others don't.

2. At this stage, no matter what door the contestant chooses, the probability is $\frac{1}{3}$ that she will pick the door with the desirable prize.

3. Having announced her choice, but before the door is opened to disclose the prize, Monty Hall, the host of the game show, opens one of the two doors she did not choose, revealing an undesirable prize, and offers her the chance to change her choice. Should she change?

4. Yes, she should change: The probability that her original choice is the desirable prize is only $\frac{1}{3}$, while the probability is $\frac{2}{3}$ that the other unopened door has the good prize.

5. The validity of the above answer assumes that the host knows which door conceals the desirable prize and never opens it.

B. Here is a variation on the Monty Hall problem.

1. If there are 1,000,000 doors, the contestant's initial guess has a $\frac{1}{1,000,000}$ chance of being right and a $\frac{999,999}{1,000,000}$ chance of being wrong.

2. Let's say Monty Hall opens 999,998 other doors and leaves one closed besides the one the contestant selected.

3. The probability is $\frac{999,999}{1,000,000}$ that the prize is behind the other remaining closed door, so the contestant should definitely switch.

C. And here is another variation of the Monty Hall problem:

1. Let's say there are five doors; the contestant's initial guess has a $\frac{4}{5}$ chance of being wrong.

2. Suppose Monty Hall then opens two of the losing doors and offers the contestant the chance to pick one of the other two remaining closed doors.

3. The probability is $\frac{4}{5}$ that the prize is behind one of the two closed doors other than the door originally selected.

4. Switching to one of the other doors gives the contestant a $\frac{2}{5}$ chance of winning, while sticking with the original choice gives her a $\frac{1}{5}$ chance.

III. Our next example is a problem from tennis: If the score in a tennis game gets to deuce, what is the probability of the server winning the game?

   A. Deuce occurs when the game is tied and one player has to get ahead by two points to win.

   B. It appears that this is an infinite problem because there is no theoretical limit to the number of deuces in a game.

   C. In fact, this problem can be resolved by a clever strategy.

      1. Suppose the server has a 0.6 probability of winning each point and the receiver, a probability of 0.4 of winning.

      2. The probability of the server winning the next two points is $0.6 \times 0.6 = 0.36$.

      3. The probability of returning to deuce is $0.6 \times 0.4 + 0.4 \times 0.6 = 0.48$.

   D. Let p be the probability that the server eventually wins.

      1. Either the server could win in two points (0.36) or the game could return to deuce (0.48), followed by the server's eventually winning. We get the following equation: $p = 0.36 + 0.48p$.

      2. Solving the equation, we see that the server will win with a 0.69 probability.

IV. Here is a final problem to ponder:

   A. Suppose you have three sock drawers.

      1. In one drawer, you have two blue socks.

      2. In a second drawer, you have two red socks.

      3. In a third drawer, you have one red and one blue sock.

   B. You randomly choose a drawer, reach in, and pick out a sock without looking. You see it is red.

   C. What is the probability that the other sock in the drawer is also red?

   D. Answer: You are equally likely to have chosen any one of the three red socks. For two of them, the other sock is red; for the third, the other sock is blue. Thus, the probability of the other sock being red is $\frac{2}{3}$.

Readings:

Edward B. Burger and Michael Starbird, *The Heart of Mathematics: An invitation to effective thinking*, 2nd ed.

Edward B. Burger and Michael Starbird, *Coincidences, Chaos, and All That Math Jazz: Making Light of Weighty Ideas.*

Questions to Consider:

1. Suppose I am in a room with 49 other people. What is the probability that someone in the room has the same birthday as I do? Hint: This question requires a different calculation from the one presented in the lecture. To see why, suppose that my birthday is, for example, July 10.

2. Suppose in the *Let's Make a Deal*[R] show that Monty Hall did not know the location of the big prize, and he sometimes would open the big prize door by accident. Now analyze the situation in which the contestant selects a door, Monty Hall opens another door, and it happens to reveal a worthless prize. Is the contestant better off switching, or in this case, are the probabilities for switching and sticking the same?

Lecture Nine—Transcript
Probability Surprises

No course on probability could possibly be complete without a discussion of two of the most famous examples of counterintuitive probabilistic scenarios. The first one we're going to do is the birthday problem, and then we're going to do the *Let's Make a Deal*[R] Monty Hall question. We'll start with the birthday problem.

Here's a question for you. You go into a room, and there are 50 random people in the room. Now, you know that every one of these people celebrates a birthday on some day of the year, so they have 366 possible birthdays on which to celebrate their birthday. What's the probability that out these 366 possible birthdays, that two people in the room will just by accident have the same birthday—will celebrate their birthday on the same day of the year? Well, at first it seems that that's rather unlikely. There are 366 days and only 50 people in the room. What are the chances that two of them would have the same birthday?

Well, I often teach classes that have about 50 people in them, and when I do this, the very first day of class before the students know me very well, I make a bet to them, and I say, "I'll tell you what I'm going to do. I don't know you, but I'll make you an even bet. I'll bet $1, and I want somebody to come down and bet $1, and we'll just make it an even money bet. If there are two people in the room with the same birthday, then I win your $1, and if there aren't, then you can have my $1." And you see I'm showing them how generous I am, because I'm giving them $1, when there are only 50 people in the room out of 366; how generous can you possibly be?

I might say that later in the semester this doesn't work so well, because after a few bets they discover they always lose the bets and then they change their mind; but it works the first day, and maybe you can earn some money on this yourself if you try this, because it turns out that if you have 50 people in a room, that the probability that two of them will have the same birthday, instead of being a rather rare occurrence, actually happens 97% of the time. It's almost certain that two people, randomly selected, that two people will have the same birthday—not treating people randomly, but among those 50, two of them will have the same birthday. Well, why in the world would it be the case that with only 50 people in a room, the probability of there having at least one pair of people with the same birthday is so high—97%?

Okay, so what we need to do is to think, and to figure out exactly how to compute that probability. Well, the first principle that we're going to use is the fact that either two people in the room or more have the same birthday, or everybody in the room has a different birthday. Those are the two possibilities, right? Either at least two have the same or they're all different. It turns out that the challenge of proving that all 50 have different birth dates is a much easier problem to do. So what we're going to really compute is the probability that all 50 birth dates are different, and then 1 minus that gives the probability that at least two are the same, because that's the alternative.

Remember, on the very first lecture we talked about that—that one of those two things is going to happen, and if we know the probability that they're all different is something, then the probability there's at least two the same is 1 minus that.

How do we compute the probability that all the people in the room have different birth dates? Well, this is the way we do it. We take the first person. We bring them up, we say, "What's your birthday?" And they give us their date; you know, April 3; and then we say, "Okay, we'll take the next person in the room, and ask ourselves the question, 'What's the probability that that next person has a different birth date from this first person?'" Well, the answer is that there are 365 days of possible birth dates. We're going to put February 29 in there as a possible birth date, and assume that they're equally likely, even though, of course, February 29 isn't really, but for the sake of argument, we'll assume all 366 possible days are equally likely.

What's the probability that this second person has a different birth date than the first person? Well, the answer is 365 out of 366. Now, having accomplished those two being different, what's the probability that the third person has a birth date different from both of those other two? Well, this third person can't have either the first birth date, April 3, or that second birth date, which let's say is March 2. Let's suppose that those are the two birth dates we got. Well the third person has to have a different birth date, so the third person's possibility is only 364 out of 366. Then we take the fourth person. What's the probability that that fourth person is going to have a different birth date from all of these other three? Well they have a

©2006 The Teaching Company

smaller number of possible birth dates to choose from that would be different; their birth dates have to be among 363 unused birth dates divided by 366.

The product of all of those probabilities, because all of them have to be true, so 365 out of 366 times the second person is okay. And then the probability is only 364 out of 366 that the third one is okay, that is, different from the first two; and only 363 over 366 that the fourth one is okay; and you work your way down to the 50th one. Now, the last ones in the list have probabilities of only 319 over 366, and then the second to last 318 out of 366, and the last one 317 out of 366. So although each one individually is very likely to have a birthday different from everybody else, the product of all of those numbers is going to get very small, and in fact the product—if you just multiply them out, just take a calculator and multiply them out—you'll see that that product of all those 50 numbers, 49 numbers, is .03—50 numbers if the first number is viewed as 366/366, by the way.

The probability of their being all different is the product, which is just .03. It's a very small number. And therefore the probability that two people have the same birth date is 1 minus that, which is 97%. Now, this seems sort of counterintuitive, but it's true, and you could actually verify it. The fun thing is that you can verify it. Just think about groups of 50 people, and just see that almost always they have, in fact, at least two people have the same birth date.

For example, let's look at the presidents of the United States. Now there aren't quite 50 of them, but if you look at all the different presidents of the United States, there are pairs of them that have the same birth date; namely, Polk and Harding were both born on November 2. Isn't that interesting?

Then we can look at death dates. It turns out that Fillmore and Taft both died on March 8. And three people died on the same day: Adams, Jefferson, and Monroe all died on July 4—now, that's probably not just an accident, there's some psychology involved there, but this is an example to show you.

We could look at vice presidents. If we look at the vice presidents, there was one triple of vice presidents that all were born on the same day. Now here they are: Hannibal Hamlin—you remember, I presume, Hannibal Hamlin—I don't think somebody named Hannibal will be elected very soon in the United States, by the way—but Hannibal Hamlin was born on August 27; as was Charles G. Dawes, also born on August 27; and here's a name we actually know—Lyndon B. Johnson was born on August 27. So there it is.

Here are some other famous examples of groups of about 50, and we see these accidents. The same birthday—Best Oscar Award Winners, Ben Kingsley and Anthony Hopkins were both born on December 31. And there were same death days: Humphrey Bogart and Peter Finch both died on January 14 of their years. And of course many people are still alive, so that's even more rare.

In fact, let me just show you a chart to show you the probabilities when we have fewer people or more people in the room. If you have 90 people in the room your chance of having at least two people with the same birthday is 0.999993, so it's essentially certain if you have 90 people in the room that at least two of the will have the same birthday. You can see the chart—you're still at 99.99% for 80 people in the room; 99.9% for 70 people in the room; 99% still for 60 people in the room; 97% for 50 people in the room. When you get down to 40, it's 89%; 30 still has a 71% chance of having two the same, and the break-even point is 23 people. At 23 people you have a 50–50 chance, an even chance roughly, that at least two of them will have the same birthday. So I think this birthday phenomenon is really interesting, and it's something you can literally do at home. So this is great fun.

That is the first, and I would say the most famous counterintuitive probability example is this birthday problem, and it's great. Let's go on to the next one that is also extremely famous and really wonderful to think about. And this has to do with a game show that used to be on that was called *Let's Make a Deal*^R; and *Let's Make a Deal*^R had a host whose name was Monty Hall.

The way that *Let's Make a Deal*^R worked was this. Monty Hall would have this whole audience full of people, and they'd be dressed up in costumes. Somebody dressed up like a raisin, and somebody dressed up like a superhero, and so on. And he'd pick somebody and he'd say, "Come on down," they'd come on down and on the stage were three doors—door #1, door #2, and door #3. And Monty Hall would say, "Contestant, there are three doors here, and behind one of these doors is a beautiful pink Cadillac with these fins in the back and it's a gorgeous thing. It's probably a convertible, and it's just great. If you have that you'll be happy the rest of your life," he says

(Monty Hall). And the contestant is really excited, "Yes, that's what I want."

That's behind one of the three doors. And behind the other two doors there are other things not quite as desirable. One is a mop and one is a bucket. A bucket and mop, those are behind the two other doors, and a Cadillac. So Monty Hall says to the contestant, "Now, contestant, would you rather take the thing that's behind door #1, door #2, or door #3?" And so the contestant ponders this question. Of course, they're all closed so he doesn't know which one has it. He's pondering this question, and the audience is helpfully giving advice, "Go for one! One! Go for Two! Two! Go for Three! Three!" and so finally the contestant says, "Okay, okay, thank you, Monty Hall. Monty, I'll take door #2." And a hush falls over the audience—door #2.

So Monty Hall says, "I'll tell you what I'm going to do. Before you make this final—I know you chose door #2—but I'll tell you what I'm going to do. I'm going to open one of the other doors." And silence reigns, and Monty Hall carefully says, "I'm going to open door #3," and so Monty Hall opens door #3, and behind door #3: a bucket. Whew. Ah—a sigh of relief. But, of course, Monty Hall knows where the car is, and Monty Hall will never open the door with the car. That's against the rules. So Monty Hall never opens the door with the car; he knows where it is, so he chooses one of the doors that do not have a car.

Now there are two remaining closed doors—door #2, the original choice of the contestant, or door #1. And Monty Hall says to the contestant, "Contestant, (for the raisin, "Mr. Raisin," he'd say) would you like to stick to your original guess, door #2, or if you'd prefer you can now switch to door #1." And now, once again, the audience gives helpful advice—"Switch! Switch! Stick! Stick! Switch! Switch!" and so the contestant is battling in a paroxysm of indecision—should I stick? Should I switch? Should I stick? And so the contestant says, "I'm going to go with my gut. I'm going to go with my original feeling. I'm going to stick to door #2." And so door #2 is opened, and what's behind door #2? A mop; a mop. Very sad—it's a mop.

Okay, the contestant is dejected, and so what I'm going to tell you now is that the contestant made a mistake. The contestant should have switched. You see, the contestant thought that there were two closed doors, and so it was equally likely that the car was behind either of those two closed doors; that it was a 50–50 chance, a 50% probability of being behind two closed doors, so it didn't matter. So the contestant went with his gut, that first feeling, and therefore lost the car.

Now in fact, it's not the case that the probability is 50–50. In fact, the probability is 2/3 that the car is behind the door that the contestant would get by switching—2/3, not 50–50. Let's see why. When the contestant made the original choice of three doors, the contestant had a 1/3 chance of correctly guessing the door that had the car. We all agree with that, right? There are three closed doors; randomly choose one; the probability is 1/3 that the contestant chooses the correct door.

Well, that's the same thing as saying that there's a 2/3 probability that the contestant chooses the wrong door, right? There's a 2/3 probability that the car was behind doors number #1 or #3—behind one of those two doors—but after Monty Hall opens one of those two doors, what happens? Well, if the contestant switches, the contestant will definitely be choosing the door with the car. In other words, all the contestant had to do to win the car, should the contestant choose to switch, is to have guessed wrong on the initial guess. You see? So that's a way to see that the probability of winning if you switch is 2/3.

Okay. Let's do it one more time to make sure, because this is quite counterintuitive to think about the possibility that there's a difference in probabilities, so let's think about it a different way. In fact, let me just to illustrate this to drive it home a little bit; let's do a slight variation on the Monty Hall game.

Suppose instead of just three doors, suppose Monty Hall had, say, a million doors on stage, and the car was behind one of those million doors. And the contestant comes up and Monty Hall says, "Well, what door would you choose?" And the contestant says, "Well, I would like door #623,447." So Monty Hall says, "Oh, fine. We'll provisionally hold that door, keep that door closed," and then Monty Hall opens every other door except for one other door. So he opens 999,998 doors, leaving only door number #123,247 closed.

Now, don't you feel that the chance of coming home with the car, if you were the contestant, and you randomly picked this door #623,447, if that was the door that you originally picked, you can't have felt very confident that you were going to drive home in a pretty Cadillac, because there was only one in a million chance that that would happen. But then after Monty Hall opens all of these other doors except for one, he's pointing the direction to the place that actually has the Cadillac behind it, you see. And so you're much better off switching than to stick to your original guns.

Now, in the case of the million doors it's more obvious than in the case of three, but the principle is exactly the same. As long as you miss the guess in your first guess, then if you switch, you will definitely win the Cadillac, and your probability of missing on the first guess with the three doors situation is 2 out of 3, and therefore you'll win.

Okay, so this is the famous *Let's Make a Deal*[R] problem with the Monty Hall problem; and this became famous because it appeared in the Marilyn vos Savant column a few years ago, and people wrote in and said, "Oh, I can't believe that you should switch and stick and so on, that switching makes any difference," and they wrote these long letters, and there was a big controversy. In fact, even some mathematicians wrote in claiming that it didn't matter if you stuck or switched, and so this made a splash. There was lots of controversy.

But what I'd like to do in order to make this a little bit different, to show you some variety, is let's do the same problem, but with a slight variation that might be sort of somewhat intriguing. Suppose the variation is the following: instead of having three doors, suppose that once again there's one big prize, a Cadillac, behind one of five doors this time. So we'll have five doors on stage, and there's one Cadillac behind one of the doors, and the contestant, once again, comes up and chooses one of the doors at random, but in this instance, Monty Hall chooses to open two of the unopened doors— not the one that was selected by the contestant—but two of the remaining four doors, neither one, of course, containing the Cadillac. Now, the contestant could either stick to the original guess, or the contestant could choose one of the two remaining closed doors, other than the door that was originally selected. Do you follow me?

So let's be specific. Suppose that the contestant first chose door #2, and then Monty Hall opens doors #1 and #5. So that leaves three closed doors: #2, #3, and #4. What's the probability that the contestant will drive home in a Cadillac if the contestant first, sticks to the original guess, or switches to either door #2 or to door #3 [sic door #3 or door #4]?

Well, let's think about it. By the strategy of analysis that we did before, I think we'll be able to figure this out quite clearly. So when the contestant first made the initial guess of door #2, that contestant had a 4/5 chance of being wrong. So 4/5 of the time the Cadillac will be behind the other doors, other than door #2. Now, Monty Hall opens two of the other doors, in this case door #1 and door #5. So that means that among the two remaining closed doors, other than door #2, there's a 4/5 chance that the Cadillac is behind doors #2 or #3 [sic #3 or #4], because, you see, there was a 4/5 chance that the door was something other than 2, and now Monty Hall says, "Well, it's not behind #1 or #5," basically. He's giving information when he opens up doors #1 and #5.

That means that the Cadillac is either in door #3 or #4, with probability 4/5. See, it's a little tricky. So there's a 4/5 chance that it's behind one of those two doors, so if you switch what's the probability that you will get the car? Well, there's a 2/5 chance that it's behind door #3, a 2/5 chance that it's behind door #4, and a 1/5 chance that it's behind door #2, your original guess. So you double your probability of getting the car by switching. You can do all sorts of variations on the Monty Hall problem; so this is the famous Monty Hall problem.

Okay. So I think these are delightful examples of probability questions. Now I'm going to do another one that comes up in real life, which is in playing tennis. Suppose that you are playing tennis or watching tennis, and you're watching two people play tennis, and of course the server, when the server is serving, has an advantage, because for good tennis players, serving is quite an advantage in most cases.

Suppose we get to deuce in a game, and maybe we're watching at Wimbledon or something, and the players get to deuce, and we're asking ourselves the question: What is the probability that the server will eventually win the game? Well, let's assume that when the server is serving, the server has a probability of 0.6 (60%) of winning any individual point in which the server is serving. The server has a better chance of winning than the receiver (0.6). And the question is: After you've gotten to deuce, what is the probability that the server will win the whole game?

Well, this is a rather tricky question, and the reason it's rather tricky is that there are really infinitely many things that could happen; namely, you could start at deuce; it could be that after just two points, the game ends. By the way, for those of you who don't play tennis, let me explain that when you're playing tennis, and in order to win a game of tennis, if both players get three points, then you get to what's called deuce, and you have to win the game by two points. So that means that in order to win, either the server has to win two points or the receiver has to win two points in order to be ahead by two. Or if the server wins one and then the receiver wins one, then you get back to deuce; or vice versa, if the receiver wins one and then the server wins one and you get back to deuce and start again. That is, you have to continue until one of the two players wins both the points and wins the game.

The reason that this is a real challenge is because it looks like an infinite question. It's an infinite question because there are infinitely many things that could happen. You could come back to deuce once; you could come back to deuce twice; you could come back three times, four times, five times. In order to compute the probability that the server wins, it looks like you have to do infinitely many things. You'd have to first say, "What's the probability that the server wins after just two points? What's the probability that it gets to deuce, and then the server wins? What's the probability that it gets to deuce twice, and then the server wins?"

That would be an infinite series, and by the way, you could solve this problem using this infinite strategy. But there's a very clever way to do it that makes this a finite problem instead of an infinite problem, and that is the following. Here's the way you look at it: The probability that the server will win in exactly the next two points is (0.6×0.6), because remember we're assuming that the server has a 60% probability of winning any individual point; so the probability of winning two in a row is the product of winning the first one times the probability of winning the second one. So that's 0.36.

Well, what's the probability of returning to deuce? Well, there are two ways that you could return to deuce. The server could win the first point, which happens 60% of the time, or with probability 0.6, and then the receiver wins with probability 0.4. Or the receiver could win the first point, and then the server could win with 0.6. So the probability of coming back to deuce—there are two ways to do it; either the server wins, then the receiver; or the receiver, then the server—and their respective probabilities are (0.6 × 0.4), which is (0.24 + 0.4 × 0.6), which is again 0.24, for a total of 0.48. In other words, just under 50% of the time, you'll return to deuce after 2 points.

Now, here is the clever strategy. Instead of going on with an infinite series, the strategy is the following. You say, "Okay, let P be the probability that the server will eventually win the game. That is some definite number. There's some probability that the server will win. And you can break it down into two possibilities: Either the server will win in just 2 points, and that happens with probability 0.36; alternatively, the game could return to deuce, which happens with probability 0.48, and then after returning to deuce, the probability that the server wins the game is P again.

Altogether then, the probability of the server winning would have to satisfy this equation. Solving for this equation (subtracting $0.48P$ from $(1 \times P)$ gives $(0.52 \times P)$, and that is equal to 0.36; dividing by 0.52 gives a total of 69%), so means to say that the probability of the server winning is 69%. You notice that it's higher than the probability of winning any individual point. So this is an example where you can take what appears to be an infinite process and condense it down to a finite one.

Now, I think that The Teaching Company courses lack homework problems, and so I'm going to leave you in lecture with another famous probability problem for you to ponder, and that is the following thing: Suppose that you have three sock drawers; three drawers, and in the drawers you have socks. In one drawer you have two blue socks, and in the other drawer you have two red socks, and in the third drawer you have a red sock and a blue sock. Follow me? Maybe this isn't the way you'd arrange your socks, having three of each color, but okay, that's the way it works. But you don't know which one of the three drawers has those various arrangements of socks. So what you do is you stand in front of the drawer, you open a drawer, and you reach into the drawer and randomly pick out a sock, and that sock is red. Now, question: What is the probability that the other sock in the drawer is also red?

Let me summarize this one more time. Let me say it one more time, so you make sure that you understand the problem well, and you can work on this yourself—and you can simulate it if you want, by the way, by taking just like three cans and putting a red and a blue thing, and two blues, and a red and a red, and then mixing them up and reaching in. Let me just say it one more time: You've got three drawers. One drawer, but you don't know which, has two blues; one drawer, you don't know which, has two reds; and one drawer has a red and a blue. You randomly choose a drawer, reach in and pick out a red sock. What's the probability that the other sock is red?

This is a little bit surprising, and I look forward to your working on it, and I hope that you will too.

In the next lecture, we're going to look at another very famous example in probability theory that's also counterintuitive, and so I'll look forward to telling you about the problem of having two children who are boys in the next lecture. I'll look forward to seeing you then.

Lecture Ten
Conundrums of Conditional Probability

Scope:

An important concept used to help us find our way through probabilistic complexity is the idea of *conditional probability*. Conditional probability refers to a situation in which we begin with a clear probabilistic scenario but are then told more information. The additional information alters the probabilities, but frequently, the change is challenging to analyze. Principles of dealing correctly with conditional probability can guide us to correct answers, but these are tricky and highly non-intuitive issues. The famous *Bayes' theorem* describes the relationships among related conditional probabilities. The ideas of conditional probability are introduced via some probabilistic conundrums that delightfully puzzle us.

Outline

I. As our knowledge and information about possibilities in a situation change, the probabilities of events change. This concept is called *conditional probability*, the topic of this lecture.

II. To introduce conditional probability, we will consider a collection of 27 cards that have been chosen to illustrate the idea. There are 21 black cards, of which 9 are face cards, and 6 red cards, of which 3 are face cards.

    A. We can answer questions about the probability of choosing a certain type of card from this group of cards.

        1. What is the probability of choosing a face card? Because we have 12 face cards, the answer is $\frac{12}{27}$.

        2. What is the probability of choosing a red card? Because we have 6 red cards, the answer is $\frac{6}{27}$.

    B. Conditional probability enters the picture when we are told one of the characteristics that cuts down the population.

        1. For example, what is the probability of getting a red card given that we have chosen a face card?

2. There are 3 red cards out of the total 12 face cards; thus, the conditional probability of choosing a red card given that we have chosen a face card is $\dfrac{3}{12}$.

III. Let's look at another question that relates two different conditional probabilities.

   A. What is the probability of getting a face card that is red?

      1. This question involves two probabilities: the probability of choosing a face card and the conditional probability of choosing a red card given that we have a face card.

      2. The answer is the product of two probabilities: the probability of choosing a face card from among 27 cards $\left(\dfrac{12}{27}\right)$ times the conditional probability of choosing a red card given that we have a face card $\left(\dfrac{3}{12}\right)$, or

$$\frac{12}{27}\times\frac{3}{12}=\frac{1}{9}.$$

   B. We can look at the same situation backward.

      1. What is the probability of getting a red card that is a face card?

      2. The analysis is the same: the probability of choosing a red card from among 27 cards $\left(\dfrac{6}{27}\right)$ times the conditional probability of choosing a face card given that we have a red card $\left(\dfrac{3}{6}\right)$, or $\dfrac{6}{27}\times\dfrac{3}{6}=\dfrac{1}{9}$.

IV. Bayes' theorem is a principal tool that is used to deal with conditional probability.

   A. Suppose A represents one characteristic (such as "red card") and B represents another characteristic (such as "face card").

   B. Bayes' theorem relates two conditional probabilities, the probability of A given B and the probability of B given A.

   C. It can be presented in two ways:

$$P[B] \times P[A|B] = P[A] \times P[B|A] \text{ or } P[A|B] = \frac{P[B|A]P[A]}{P[B]}.$$

V. Conditional probability can surprise us. Consider the following scenario: Suppose you meet a man and learn that he has exactly two children.

    A. Suppose that you learn that his older child is a boy.

        1. Therefore, we know that two of four possibilities are eliminated (two girls [GG] or an older girl and a younger boy [GB]), leaving the possibility that he has two boys (BB) or an older boy and a younger girl (BG).

        2. Of the remaining two equally likely possibilities, one is boy-boy. Thus, the probability that both children are boys given that the older child is a boy is $\frac{1}{2}$.

        3. This is called conditional probability.

    B. But suppose you ask the man instead, "Do you have a son?" and he answers, "Yes."

        1. The GG possibility is eliminated, and three possibilities remain, GB, BG, and BB.

        2. Thus, the probability that both of his children are boys given the knowledge (or "condition") that at least one is a boy is $\frac{1}{3}$.

        3. Notice that the answer is not $\frac{1}{2}$. The information that at least one child is a boy affects the probability differently than the information that the older child is a boy.

    C. Suppose you had asked the following question of the man instead: "Do you have a son who was born on a Tuesday?" and he answers, "Yes."

        1. Most people's intuition is that this birthday information is irrelevant and should yield the same probability as the previous version of the problem.

        2. To do this calculation, we begin by writing down all the possible day-of-the week and gender combinations, and we find that there are 196 in all.

        3. We then narrow down the possibilities by focusing on the pairs for which at least one child is a boy born on a

Tuesday. We find we have 13 BB possibilities, 7 BG possibilities, 7 GB possibilities, and of course, no GG possibilities, for a total of 27.

4. The probability that both children are boys given that at least one is a boy born on a Tuesday is $\frac{13}{27}$, which is between $\frac{1}{3}$ and $\frac{1}{2}$.

VI. Let's look at another problem: Suppose we have two urns, each containing 10 balls. In one urn, we have 7 blue and 3 red balls, and in the other, we have 3 blue and 7 red balls. We can't tell which urn is which.

A. I select an urn at random and draw a red ball from it; then I put the ball back in the urn and choose a ball again from the same urn, and it is red. I choose a third time and get a red ball and a fourth time and get a red ball.

B. What is the probability that the urn I chose was the one with 7 red and 3 blue balls?

1. One strategy might be to imagine having 20,000 people performing the same experiment, randomly choosing one of the two urns and randomly drawing out 1 of the 10 balls four different times.

2. Logically, about half the people (10,000) would choose the blue-heavy urn and half, the red-heavy urn.

3. Out of the 10,000 people who chose the red-heavy urn, how many would we expect to choose red balls four times in a row?

4. Each of the four times one of the people reaches into the red-heavy urn, he or she has a 70% chance of getting a red ball. Therefore, we arrive at this equation: $0.7 \times 0.7 \times 0.7 \times 0.7 = 0.2401$, or 2,401 of the 10,000 people.

5. However, for the blue-heavy urn, people have only a 30% chance of getting a red ball each of the four times a ball is chosen. The equation is: $0.3 \times 0.3 \times 0.3 \times 0.3 = 0.0081$, or 81 of the 10,000 people.

6. We know, then, that 2,482 people would draw four red balls. Therefore, the probability that the person choosing

four reds is drawing from the red-heavy urn is $\dfrac{2,401}{2,482}$, or 97%.

C. Let's change the scenario a little and draw a ball out of an urn 50 times instead of 4 times.

    1. Let's say we find that 27 times, we choose a red ball, and 23 times, we choose blue. What is the probability that we are choosing from the red-heavy urn?

    2. Surprisingly, our calculations show us that the probability is again 97%!

Readings:

Edward B. Burger and Michael Starbird, *The Heart of Mathematics: An invitation to effective thinking*, 2nd ed.

Peter G. Moore, *The Business of Risk*.

Jeffrey S. Rosenthal, *Struck by Lightning: The Curious World of Probabilities*.

Questions to Consider:

1. Someone tells you the following: "I met a man who told me that he has exactly two children. I asked him one question, but I can't remember what question I asked. It was either 'Is your older child a boy?' or 'Is your younger child a boy?' I remember that he answered yes." What is the probability that both of the man's children are boys?

2. Suppose you have two urns, one of which contains 10 red balls and the other, 5 red balls and 5 blue balls. You select an urn at random and draw a red ball from it; then you put the ball back in the urn and choose a ball again from the same urn, and it is red. You choose a third time and get a red ball and a fourth time and get a red ball. What is the probability that you are reaching into the red urn?

Lecture Ten—Transcript
Conundrums of Conditional Probability

Welcome back. In this lecture we're going to introduce a very basic concept of probability that's associated with what happens when we're asked a probabilistic question, but then we're given more information. It changes the probability because we put ourselves in a more restricted arena of possibilities.

Let me explain this concept: the concept that we're going to aim for is called *conditional probability*. I'll explain it by first looking at a very specific example. Suppose that we have a collection of cards; and here we see a collection of cards, there are a total of 27 cards in this collection, and suppose that these are randomly mixed up, and we're asking the question of what's the probability of choosing a card of a certain type?

Well, we're familiar with this. We know exactly what to do. If it's equally likely that we choose any card, we can answer such questions as: What is the probability of choosing a face card? And the answer is: Well, we just count up how many face cards there are—there are a total of 12 (3 + 3 + 3 + 3)—that's a total of 12 face cards out of the total of 27 cards, so the probability of randomly selecting a face card is going to be 12 out of 27. No problem.

Likewise the probability of choosing a red card is the total number of reds that there are (6) out of the total number of cards that there are (27). So that's the probability of choosing a red card. Well, conditional probability comes in when we're told some feature of what it is that we get.

For example, suppose that we're asking the question: What is the probability of choosing a red card given that we have chosen a face card? So in other words, we have these cards, we randomly choose one, and maybe somebody else is looking at it, and says, "Oh, it's a face card," and then you ask the question: What's the probability that it's red? Well, all that does is it puts us into this collection of cards instead of the whole group of cards.

We say, "What's the probability of having a red card if we know we have a face card?" Well, there are 3 red cards out of the total of 12 face cards, so the conditional probability of having a red card, given that you have a face card, is 3 out of 12. That's conditional probability, and there's a notation for it; the notation is the probability of getting a red card and then a vertical bar says given that you have a face card. This is the basic concept of conditional probability, and it turns out to come up in all sorts of arenas.

Let me point out some facts about this conditional probability; namely, the following. Suppose that you're asking the question: What's the probability that you get a face card that is red? So among that same group of 27 cards, what's the probability that you get a face card that's red? Well, the answer is that that probability is the probability of first choosing a face card, and then among those face cards, choosing a red one—so it's the product of 2 probabilities; the probability of choosing a face card from among the whole group of 27 cards times the conditional probability of choosing a red card, given that we have a face card.

Now, let's just look at this picture to see what we mean by this. The probability of getting a face card is 12 out of 27, and I'm assuming that we're just picking one card randomly from this group of 27 cards—this specific group of 27 cards. Well, the probability of getting a face card is 12 out of 27, since there are 12 face cards. Then the probability of getting a red card, given that we're in this group of 12 face cards, is 3 out of 12. So that's the conditional probability of getting a red card, given that we have a face card—is 3 out of 12.

That means that the probability of getting a face card that is red— that is, we choose a card at random from the whole group, the probability that it's both a face card and it's red is the probability that we get a face card (12 out of 27), times the probability that we have a red card, given that we have a face card. And that product is 1 out of 9—or 3 out of 27. And you see it is the correct answer. The probability of picking a red face card is the probability of picking one of these 3 cards, which are the 3 red face cards, from the total of 27 cards.

Well, of course, we can look at the same situation backwards. In other words, we can say, once again asking the question: What's the probability of getting a face card that's red? Well, we can say that that's also the same as: What's the probability of choosing a red card from the whole group, and then within it, what's the probability of having a face card if we already know that the card we've selected is red? This is illustrated in this diagram. The probability of choosing a red card is 6—there are 6 red cards—out of the total of 27 cards, so the probability of choosing a red card from the whole group is 6 out of 27. Once we know we're in the group of red cards, the probability of getting a face card is 3 out of the 6 red cards. So the probability of getting a face card, given that we already know it's red, is 3 out of 6.

So another computation for getting a red face card is 6/27 (that's the probability of choosing a red card) × the conditional probability of choosing a face card, given that we have a red card (3/6). And once again, of course, we get the same answer of 1/9.

Now, a principle tool that is used in dealing with conditional probability is called Bayes' theorem, and what it does is it relates two different conditional probabilities such as the two conditional probabilities that we've been talking about. We talked about two different conditional probabilities; namely, the conditional probability of choosing a red card given that it's a face card; and the opposite, the conditional probability of choosing a face card given that it's red card. We saw that there was a relationship between the two, and let's go ahead and summarize it in Bayes' theorem. That's what Bayes' theorem summarizes.

It says the following: Suppose that you have two different characteristics (A and B) in some group of things—and by the way, when you're thinking about Bayes' theorem, simply think that A means red card and B means face card. Then what Bayes' theorem says is that the probability of condition B times the probability of A given B—the conditional probability of A given B—is equal to the probability of A times the probability of B given A.

Now, by the way, when I read that it's just sort of words, you know. But if you think of it as to say "red" and "face card," and you realize that what you're trying to compute is the probability that you get a card that's both red and a face card, then you can make sense of both sides of this equation and see that they're the same. Because, you see, what this says is that the probability of getting a face card—so, being among the face cards, those 12 face cards, that has a probability, 12 out of 27 in our example—times the conditional probability of getting a red card, given that it's a face card, that product is giving the answer of the probability of selecting a red face card.

But likewise, it's of course the same answer as first seeing what the probability of getting a red card is times the probability among the red cards of having a face card—which is the conditional probability of having a face card, given that it's already red. This is Bayes' theorem, a very famous theorem in conditional probability; and it can be expressed by simply dividing through by the probability of B, this is the form that it's often phrased in because it puts the conditional probability of A given B in terms of the other three probabilities.

This is an example of Bayes' theorem that talks about conditional probability, but I want to give you some examples that illustrate this concept. So let's start with a scenario that I refer to as the reunion scenario, because it can be viewed as the following thing: Suppose you go to your 25th college reunion. You meet this person at the reunion and you say, "I understand you have two children." And they say, "Yes, I have two children." And then you say, "I understand that your older child is a boy." And they say, "Yes." What is the probability that both children are boys, that that person has two boys, given the information that the first one is a boy—the older one is a boy?

Well, here's the way you think about it. As soon as you said that this person has two children, there were four equally likely possibilities for those children. Either the older one was a boy and the younger one was a boy; or the older one was a boy, and the younger was a girl; or the older was a girl, the younger was a boy; or they were both girls. Those are equally likely scenarios for the four children if you had no further information other than that they had two children.

Now we add other information: namely, that the older child is a boy. Well, that eliminates the last two possibilities, and we have only two of these equally likely possibilities remaining. One of them is that the person has two boys, and so the probability that the person has two boys is 1/2. Great—nothing surprising there.

Let's do the same situation with a slight variation. The slight variation is the following. You go up to this person and you say, "I understand you have two children." They say, "Yes, I do." And then you say, "I understand that one of your children is a boy," but you don't know which, you don't know whether it's the older or the younger, one of them is a boy. Now you ask the question: What's the probability that they have two boys? Well, in this circumstance it is a little bit different. You see, the difference is you didn't say that the older one was a boy; you just said that you have at least one boy.

Well, look at what happens. Among the four possibilities that you began with when you knew that they had just two children but had no further information, only one of those is eliminated by the knowledge that at least one of the children is a boy. Therefore, there are three equally likely possibilities remaining, of which only one is that both children are boys. So the probability of having two sons, given the information that at least one is a boy, is only 1 out of 3.

Now this is often surprising to people. It's very counterintuitive because you'd think that knowing that they have one boy, well the other one is either a boy or a girl, so why shouldn't it be 50–50? But you see this analysis shows us that the probability is actually just 1 out of 3. Well, it turns out that this particular probability paradox is actually quite subtle, and in fact it's confusing to a lot of people, and the reason that it's confusing is because some of the subtleties are things arise that make it more challenging than you would think.

I personally made a mistake about this particular problem when I tried to phrase it in a slightly different way, and it turned out it changed the probability in a subtle way that I didn't notice at first, and somebody pointed it out to me. So I got tripped up on it myself. This is really rather subtle, and I'm going to show you how subtle it is by doing a variation of this problem that I'll think you'll find it really rather astounding, at least I found it so.

Suppose we make the following just small variation on the problem. You come up to this person and you say to them, "You have two children." You've learned this someplace that they have two children. And they say, "Yes, I have two children." And then you say, "I understand that one of your children is a boy who was born on a Tuesday." And the person says, "Yes, that's correct. One of my children was a boy who was born on a Tuesday." And now you ask the question: What is the probability that that person has two boys?

Now of the all the irrelevant things that I can think of, having the child born on a Tuesday appears to be among the most irrelevant. Why would that have anything to do with this probability scenario? Well, it turns out that it alters the probability significantly, and here is the analysis. And I just find this mind-boggling, and so here we go.

Our strategy for computing the probability, as always, is to try to write down equally likely possibilities of these two children that are relevant to the information that we have. In this case, there are actually 196 equally likely possibilities. When we think about day of the week being born and male and femaleness of each of the two children in the following sense: There are 49 equally likely possibilities for the person having two boys, because the older boy could have been born on any day of the week—Sunday, Monday, Tuesday, Wednesday, Thursday, Friday, Saturday. Likewise, the younger one could have been born on any one of those seven days of the week, for a total of 49 different combinations of older boy and day of birth, and younger one and day of birth. Oh, and by the way, let me point out something that I'm assuming, which is that the day of the birth is random—that it's equally likely to be born on any of the seven days. Now, incidentally, that's not true because hospitals try to get people to be born during the weekdays, so my guess is it's actually more probable to be born on a weekday than the weekend, but we're forgetting that for now. We're assuming that it's equally likely.

So here we go: We have 49 boy–boy possibilities; that is, that the first square in each case indicates the day of the week on which the older boy was born, and the second square indicates the day of the week on which the second boy was born. So we have these 49 possibilities. Likewise, for the possibility of the older one being a boy and the younger a girl, we have again have 49 possibilities. The older boy could be born on any of the seven days of the week, and for each such day of the week, the younger girl could be born on any one of those seven days. Likewise, we have 49 possibilities for the older girl, younger boy possibility, and then 49 for the girl–girl possibility.

Now, what we need to do is to understand when we learn that at least one child is a boy born on a Tuesday, we need to look among all of those 196 equally likely things, and find those on which there is included a boy who was born on a Tuesday. Well, let's go ahead and do that. It's not hard. What we do is in the boy–boy scenario, there are 13 of these pairs which include at least one boy born on a Tuesday: namely, the first boy could be born on a Tuesday, so that's seven; and then the second boy could be born on a Tuesday. That would be seven more, except for the fact there's an overlap on the Tuesday–Tuesday possibility. That's a total of 13 boy–boy possibilities that include at least one born on a Tuesday.

For the boy–girl possibility, older boy, younger girl, there are seven additional cases that include a boy born on a Tuesday; and for the older girl, younger boy possibility, there are seven more instances where there is a boy born on a Tuesday; of course none in the girl–girl section. So just adding those up, we have a total of (13 + 7 + 7), which is a total of 27 possibilities that include a boy born on a Tuesday, of which 13 are in the boy–boy category, so the probability of having two boys, given that one of the boys is born on a Tuesday, is 13 out of 27. It's neither a half nor a third—it's something in between. Isn't that amazing? So, anyway, I find this to be a really remarkable example.

Well, I'm going to now shift gears and go on and talk about another probabilistic scenario which involves urns, and this is very important to always do something that involves urns, because in any discussion of probability, urns have to play a role, and particularly reaching in a picking a ball from an urn, that's got to be part of it. Here I have two indistinguishable urns; they look exactly the same, you see, and first of all you have to recognize them as urns. You have these two urns; they look exactly the same, and each one contains a certain number of balls, and I'll show you what they contain. They each contain 10 balls, and this one contains 7 blue and 3 red; and this one contains the opposite, this one contains 7 red and 3 blue. Okay?

Now, I have these two urns and here's the experiment that we're going to undertake. I'm going to take these two urns and just scramble them around without looking, and then I'm going to reach into one of the urns. So here's what I do. Now, I've scrambled them around so that now I don't know which one is the one that has the red dominant, red-heavy urn, and which is the blue-heavy one—red-heavy means that there are the 7 reds; that's what I mean by the red-heavy urn. So I'm trying to get some evidence about which is which. Pretend I can't look into these, by the way.

But here's the evidence that I'm going to get. I'm going to gather evidence by just reaching in and randomly choosing a ball from the urn, and picking it out and looking and seeing what color it is. Then that's going to be evidence, and we're going to evaluate how strong that evidence is for the question of whether I am reaching into the red-heavy urn or the blue-heavy urn. See, I don't know which I'm reaching into.

Here's what I do: I reach into this urn, which I've selected randomly between the two urns, I reach in and I get a red ball. Well, okay, that's something. Now I put it back in and I scramble up the balls in the urn, and I keep to the same urn now, again, and I again reach in and I pick out once again a red. And then I put it back in and I scramble it up in the same urn again. I'm always just focusing on this one urn because I'm getting evidence about this one urn. I'm trying to decide what's the probability that it's the red-heavy one or the blue-heavy. So I reach in for a third time, get red; put it back in, scramble them up, shake it up, reach in again, pick it out at random. Four times in a row I get red.

Now, question: What's the probability that I have been reaching into the red-heavy urn? You see, it's possible that I was reaching in to the blue-heavy urn because there still are 3 reds in the blue-heavy urn, and I had some probability of choosing the 4 balls that were all red from the one that only had 3. But it's more likely that I would choose 4 reds if I'm in the one that has 7 reds and 3 blues. Well, how am I going to think through what the actual probability is of these two things?

The first thing I want you to do is to ask yourself: How strong do you think this evidence is? In other words, do you think that there's a very high probability that I must be reaching into the red-heavy urn, with this evidence when I got 4 reds, or do you think that it's some evidence that it's probably red-heavy, but it's not that strong? So I want you to think about that question here.

Now, how can we think about trying to be specific about how strong this evidence is? Well, here's a strategy that's the kind of thing that we've been doing throughout this course, which is to say, let's imagine doing the experiment many, many times and seeing what we would expect. For example, suppose we had 20,000 people do this experiment; the experiment being to randomly choose one of the two urns, and then to reach in and four times in a row, pick out a ball, and then put it back in again, and randomly pick out another, and put it in, and so on. What would we expect to happen if we did this 20,000 times with 20,000 people?

Well, let's just think it through. On average, 10,000 people would be reaching into the urn that has 7 reds and 3 blues, and 10,000 people would be reaching into the other one—the one that has 7 blues and 3 reds, the opposite. How many people who reach into the urn that has

7 reds, out of 10,000 people, how many of those would we expect to choose all 4 reds, 4 times in a row to choose red.?

Well, we can figure this out. The answer is that every time we choose a ball from the red-heavy urn, we have a 70% chance of choosing a red ball. The probability on anyone reaching into the urn of getting a red is 0.7, so the probability of reaching in on 4 consecutive times and getting a red each time is $(0.7 \times 0.7 \times 0.7 \times 0.7)$, which is 0.2401. So if we conduct this experiment 10,000 times, we would expect that 2,401 of those people would have chosen 4 reds if they're reaching into the red-heavy urn.

Well, we can do the same kind of analysis for the blue-heavy urn. Suppose that we're reaching into the blue-heavy urn, it's certainly possible that we could choose a red each of 4 times; but it's probability is much less, since the probability of choosing a red ball is only 0.3 for each individual instance; so the probability of choosing 4 in a row is $(0.3 \times 0.3 \times 0.3 \times 0.3)$, which is only 81 out of 10,000.

Now if we return to our scenario of having 10,000 people reach into this urn that's red-heavy and 10,000 reaching into the urn that's blue-heavy, we expect that 2,401 will draw 4 red balls from the red-heavy one, and only 81 will draw 4 red ones from the blue-heavy one, out of the 10,000. That's what the probability would dictate. So that means that in total, among this whole 20,000 people—10,000 of whom picked from this urn, 10,000 from this urn—a total 2,482 people will draw 4 red balls. Of those, 2,401 are dealing with the urn that is red-heavy, that has the 7 reds in it; and only 81 are dealing with the urn that is blue-heavy. So in order to compute the probability, it's just $(2401 \div 2482)$. It's 97%. So the evidence is very strong, if you get 4 reds that you're in the red-heavy urn.

Well, that may or may not be surprising to you, but what I would like to do is pose a slight variation on the question, and see if I can successfully surprise you. Here is a slight variation on it. We have the same situation—we have these two urns, one has 7 reds, 3 blues; one has 7 blues, 3 reds. We scramble them up randomly, and reach into one of the urns and start choosing balls. Sometimes we get red ones; sometimes we get blue ones. But this time instead of just reaching in 4 times, suppose we reach in 50 times—each time we pick out a ball, put it back in, scramble them up; reach in again. And among those 50 times, 27 times we get red and 23 times we get blue.

Here's a question, and I pose this to you as a thought question: What do you think that tells us about the probability that we're in the red-heavy urn? If we got 27 reds and 23 blues, how strongly do you think that tells you that you're in the red-heavy urn? In other words, do you think, "Well, yeah, it's more likely to be red-heavy, but just a little bit." Most people I've talked to, their intuition tells them that there's just a little bit of a chance higher that you're in the red-heavy urn, because those two numbers are so close—27 to 23—27 reds to 23 blues. So it seems like the evidence is, for example, not as strong as the evidence of just picking 4 in a row that were all red.

Well, the possibly surprising answer is that, in fact, the probability is exactly the same that you are in the red-heavy urn. The probability if you choose 27 reds and 23 blues, that evidence is exactly the same as just choosing 4 and having them all be red; and you can compute this probability in a similar way that justifies it, by looking at the probability of being in the red-heavy urn and choosing 27 reds and 23 blues, and this denominator is the total number of the probability of choosing 27 reds and 23 blues from either of the urns. Computing that, it turns out everything cancels and you end up with just exactly the same probability formula that we would have written down for the case of just picking 4.

I hope that you are at least a little bit surprised by these examples of unusual probability and particular conditional probability. In the next lecture we're going to talk about the very interesting concept of probability as a measure of belief. I look forward to seeing you then.

Lecture Eleven
Believe It or Not—Bayesian Probability

Scope:

What does probability mean in the real world? Probabilists do not agree. Mostly in these lectures, we've focused on the frequentist view of probability; namely, that if we repeat an experiment in question many times, the percentage of successful outcomes is the probability. However, another view of probability is that it measures a person's belief in the likelihood of the item in question. "Did Shakespeare write *Hamlet*?" We can't do a repeatable experiment pertinent to this question. A frequentist holds the view that probability applies only to experiments whose outcomes are random and, therefore, would not discuss the *Hamlet* question as one susceptible to probabilistic comment. Bayesian probability concerns itself with describing a weighted assessment of possibilities, then develops a method for revising that assessment as more evidence is amassed. The different views of probability are intriguing to consider, and in some cases, adopting one philosophy or another has practical implications.

Outline

I. In most of the examples in earlier lectures, probability could be interpreted as the fraction of successes in a series of identical experiments or trials.

    A. An example would be saying that the probability of rolling a die and coming up with a four is $\frac{1}{6}$. That is, if you rolled the die many times, about $\frac{1}{6}$ of those times would show four.

    B. This view of probability is called *frequentist probability*.

II. Another use of probability is to express quantitatively our degree of belief in some statement.

    A. For example, if I say that the probability is 98% that Shakespeare wrote *Hamlet*, you'll know that I believe very strongly that Shakespeare was the one who indeed wrote Hamlet, but that there is some small possibility that someone else was the actual author of the play.

    1. Saying that we believe there is 98% probability that Shakespeare wrote *Hamlet* makes a statement about the strength in our belief.

    2. But this does not mean that if 100 Shakespeares were born, 98 of them would have written *Hamlet*.

B. We have two kinds of situations in which we use the same word—*probability*.

C. As a measure of the strength of belief, *probability* expresses our uncertainty, but the two kinds of probability are different kinds of things.

III. When probability is used to express quantitatively a degree of belief, it must be clear what all the possibilities are.

A. Among the various potential states of the world, we express the relative probabilities of those different states being the correct one.

B. And we assign to each such possible state of the world a probability that that one is the correct one

C. The sum of the probabilities is 100%.

IV. To ground our discussion, let's take an example of fish in a stream.

A. We're interested in what fraction of fish in the stream are trout, from the possibilities of 5%, 15%, 25%, …, 95%.

B. Before any data are collected, we assume that we have no bias; we establish the 10 possibilities and give each the same probability.

Hypothesis: This percentage of fish in the stream are trout	Probability of this hypothesis
5%	0.10
15%	0.10
25%	0.10
35%	0.10
45%	0.10
55%	0.10

65%	0.10
75%	0.10
85%	0.10
95%	0.10

C. Suppose we catch three fish—trout/trout/non-trout. We would naturally believe that it is more likely that the percentage of trout is high.

D. We can update our probabilities for the various potential percentages of trout in a stream by doing a thought experiment in which we imagine 10,000,000 fishermen—1,000,000 fishing in each of 10 different universes (one for each hypothesis).

    1. We can calculate how many of those fishermen would catch a trout/trout/non-trout combination in their respective streams.

    2. The following table shows our calculations:

Hypothesis: If this percentage of fish in the stream are trout	Then of 1,000,000 fishermen, this many catch two trout, then one non-trout
5%	2,375
15%	19,125
25%	46,875
35%	79,625
45%	111,375
55%	136,125
65%	147,875
75%	140,625
85%	108,375
95%	45,125
Total	837,500 in all streams

E. We can now update our belief system.

    1. Our *a priori* assumption, before catching any fish, was that each hypothesis is equally likely.

    2. Then, we caught two trout, then one non-trout.

3. We can now recalculate the probability that the stream contains 5% trout by dividing 2,375 by 837,500. Likewise, we can recalculate the probability that it is a 15% stream by dividing 19,125 by 837,500, and so forth.

4. By applying Bayes' theorem to make an update to our previous resulting distribution, we get a new distribution that has most of the probability concentrated in the choices 35%, 45%, 55%, 65%, and 75%.

F. If we catch another trout and another non-trout, we can perform the same type of calculations using our new, updated distribution.

1. Now we have evidence that changes our sense of the possibility; we have, for example, many more fishermen in the 65% stream than in the 5% stream.

2. As we catch more fish, the evidence will dominate over our initial estimate, thus reflecting the Law of Large Numbers.

3. After catching 100 fish, we have a very strong belief that we have a 65% stream, but about a 10% chance that it is a 55% stream or a 10% chance that it is a 75% stream.

V. Thus, we have two views of probability:

A. The frequentist probability is the view in which probability is defined in terms of long-run frequency or proportion in outcomes of repeated experiments.

B. Bayesian probability is the view in which probability is interpreted as a measure of degree of belief. In this view, the concept of probability distribution is applied to a feature of a population to indicate one's belief about possible values of that feature.

VI. Let's look at another example of updating our probability distribution in the field of medicine.

A. A doctor narrows a patient's illness to three possibilities: A, B, or C.

B. After assessing the patient, the doctor assigns probabilities of the patient having the diseases as follows: A: 50%, B: 40%, C: 10%.

C. After a more thorough exam, a symptom, S, is discovered, and the doctor knows what the probability is of a patient with each of the diseases exhibiting this symptom.

D. Therefore, we update our initial probability distribution:

Hypothesis: Patient has this disease	Probability of this hypothesis	Number of imagined patients[1]	Probability of showing S	Number of patients showing S[2]	Updated probability of this hypothesis
A	50%	5,000	10%	500	500/2,500 = 20%
B	40%	4,000	30%	1,200	1,200/2,500 = 48%
C	10%	1,000	80%	800	800/2,500 = 32%

[1] Out of 10,000
[2] Note that the total is 2,500.

E. We see that B is now the most probable disease, replacing disease A.

Readings:

Donald A. Berry, *Statistics: A Bayesian Perspective.*

E. T. Jaynes, *Probability Theory: The Logic of Science.*

Questions to Consider:

1. Bayesian probability involves having an a priori distribution and updating it in light of evidence. What is the influence of different a priori beliefs after a great deal of evidence is accumulated? Why?

2. Suppose your a priori belief about a coin is that you are 100% certain that it will always land heads. You flip the coin and it lands tails. Then you cannot update your probability distribution because you ascribed 0 to the probability of ever getting a tail. What went wrong?

Lecture Eleven—Transcript
Believe It or Not—Bayesian Probability

Welcome back. In most of the earlier lectures we have been thinking about probability as interpreted as a fraction of successes in a series of identical experiments or trials that we undertake, and a paradigmatic example has been that we think about taking a die, and we roll it, and we ask the question: What's the probability that the die will come out to be a 4? And it's 1 out of 6 because if we roll the die many times about 1/6 of those times the die would come out to be a 4.

Well, this basic view of probability is called the frequentist probability, because it's talking about the frequency with which a repeated event happens. But there's another sense in which we often think in terms of probability that measures really a quite different kind of phenomenon, and so we sometimes wish to use probability to express in some sort of a quantitative way the degree to which we believe something.

For example, I'll give you an example that might make this a little bit clearer. Suppose we are thinking about who wrote *Hamlet*. Now, of course, most people think, and most of us believe that Shakespeare wrote *Hamlet*, but we've heard the stories about the possibility that somebody other than Shakespeare actually wrote the Shakespearian plays. So we might express our confidence that it's almost certain that it was Shakespeare who wrote *Hamlet* by saying there's a 98% probability that Shakespeare was the one who wrote *Hamlet*, but that there's some probability, some chance that somebody else wrote it.

Well, if I say that, then you'll have a sense of how strongly I believe that Shakespeare was indeed the one who wrote *Hamlet*, but that there is at least some possibility that someone else might have actually been the author of Shakespeare's plays. Well, when we say that there's a 98% chance that Shakespeare wrote *Hamlet*, notice that we're making a kind of a statement that is a statement about the strength of our belief of something.

It is quite a different kind of statement from the frequentist statement about the probability of rolling a 4 when we roll a die; namely, when we say there's a 98% chance that Shakespeare wrote *Hamlet*, we certainly don't mean anything like the following, we don't think to ourselves, "Well, if 100 Shakespeares were born, 98 of them would have written *Hamlet.*" We don't mean that; that's just silly. That's not what we mean. We really have two kinds of situations in which the same word, *probability*, is used; and one is a measure of the strength of our belief in something. It expresses in a quantitative way our sense of uncertainty about the different ways in which the world could be.

In this lecture we're going to be talking about taking this phenomenon of life—that is, of expressing our level of uncertainty—and we're going to make it clear. We're going to say, "Okay, what do we really mean by expressing and using probability as an expression of the degree of belief?" So let's try to be very clear on what that means. So what it means is that we have in our minds some collection of potential states of the world; and for each state of the world, we have a sense of the relative probability, the strength which with we think it might be in this state, or that state, or the other state. So, in other words, we assign to each possible state of the world a probability; that is, the probability that that state is in fact the way the world is.

And if we want to clarify this concept, we would want to make sure that the sum of the probabilities of the different alternatives adds up to 100%. If we've listed all of the possible states that the world could be, then we want our sense of belief to be distributed so that the probability is 100% that one of them happens, but we just don't know which one.

Well, let me give you a common situation in which this exact kind of thing occurs; namely, you go to a doctor, and the doctor looks at you in a cursory fashion at first, and says, "Well, given your symptoms, it could be one of three diseases (A, B, or C)," and then the doctor can say, "I don't know which of those three it could be, but I would say there's a 50% chance that you have disease A; you have a 40% chance that it's disease B; and a 10% chance that it's disease C." So you see that that statement gives a sense of the idea of what the possible states of the world are, and gives a quantitative statement about what the strength with which the different possible states of the world are actually the true state.

What we're going to do in this lecture is then try to explore the concept of how to formalize this a little bit, and then how evidence is used to change our sense of the certainty of the different potential states of reality. Okay, so let's begin this discussion, then, with a specific example to really ground our discussion.

Let's imagine that we come up to a stream, and when we arrive at this stream, we don't know anything about the fish in the stream, and let's suppose we're asking ourselves the question: About what fraction of the fish in the stream are trout? Now, before we come up to this stream, we're completely ignorant, and we say to ourselves, "I have no bias about what fraction of the fish that are in this stream are trout. It could be between 0 and 10%, it could be between 10% and 20%, 20% and 30%, 30% and 40%, and so on, with equal probability." So this is before I've collected any data, and I have complete ignorance about trout, and so on. I have no bias. I'm completely agnostic as to the percentage of trout in the stream.

Now what I'm going to do is I want to express that complete doubt by drawing a chart of what I mean by that. Here's what I'm going to say. I consider that I have 10 possible hypotheses as to the percentage of fish in the stream that are trout. And instead of saying between 0 and 10%, let's just say 5%, 15%, 25%, 35%, and so on, up to 95%. So those are the 10 possibilities that I'm considering for the percentage of trout in the stream. And I could summarize my agnosticism on this point by just saying that the probability of any one of those 10 is 10%. There's a 10% chance that there are 5% of trout in the stream; there's a 10% chance that there are 15% trout in the stream; 10% that there are 25% trout in the stream; and so on.

If I wanted to draw a chart to describe this, a histogram—the way I would describe it is this: that these are the potential states of the world; that is the potential fraction of percentages of trout in the stream, and each one I'm ascribing a probability of 10% to each one of those possible states of the world. At this stage I have a description of my sense of uncertainty, and I've specified my sense of uncertainty by saying that I have an equal view that every possibility is equally likely.

Well, now, suppose that I get some evidence. In other words, I fish. I cast the line and I catch a fish, and then I catch another fish, and I catch another fish. And suppose in doing that, I catch first a trout, then I catch another trout, and then I catch a non-trout. Well, now I have evidence, and so my opinion about the state of the world that is expressed as a probability of being a 5% stream; or a 15% stream; or a 25% stream; or a 35% stream, and by that I mean 35% of the fish in the stream are trout; my beliefs now have altered because I have some evidence. Two out of three of the fish I caught were trout, so that leads me to say, "Well, it's probably more apt to be the case that the stream contains a higher percentage of trout than a lower percentage of trout."

But can we be very specific about how that evidence alters our distribution of beliefs in these 10 possible states of the world? Well, yes we can. We can be very specific, and this is what I'm going to do, is to show you how that evidence can be used to update our distribution of our sense of uncertainty—our distribution of the probability—of the different states of the world. Let's be specific about how we can use the evidence to update our distribution, and get a new distribution of the probabilities of the various states of the world.

This is interesting, because what it is, is it's sort of specifying something that we do in everyday life. In everyday life we may very well have cases where we don't really know exactly the way the world is, but when we get evidence it alters our sense of proportion about how strongly we believe something is true, or something else is true. This is making it very precise. Okay, so let's see how we would use this evidence of catching trout, trout, non-trout to update our probabilities for the different potential percentages of trout in the stream.

Well, the way that we're going to do this is we're going to do a thought experiment, and the thought experiment consists of imagining that we have 10 different universes, one for each hypothesis. We have one universe in which this stream contains 5% trout, and then we have this other universe in which the stream contains 15% trout, and then we have another universe in which it contains 25% trout, another with 35% trout, and so on. We have 10 universes. And now we imagine that 10,000,000 fishermen are going to fish in these universes, 1,000,000 fishermen for each of these hypothetical streams.

Because what we're doing is we're imagining this, in order that we can compute the experience of all those fishermen to see how many of those fishermen would catch trout, trout, non-trout in their various streams. So let's go ahead and do the arithmetic that explains this. So here we go. So we've proportioned these 10,000,000 fishermen, 1,000,000 per stream—and the reason we proportioned them that way, by the way, was because this is expressing our sense that we don't know which stream is the correct reality about the percentage of trout in the stream, and therefore, we're putting 1,000,000 for each one to make it even.

Okay. Now, here we go. Each fisherman catches three fish, and we want to know how many fishermen among these 1,000,000 fishermen in the 5% stream or lake, how many of those 1,000,000 would we expect to catch first a trout, then another trout, then a non-trout. Okay, well if you're in a 5% stream, then of those 1,000,000 fishermen, you'd expect 50,000 of them to have their first fish be a trout, because you see, in general, only 1 out of 20 fish that are caught would be a trout.

So the first fish caught by these 1,000,000 fishermen, only 50,000 of those would have caught a trout first. Of the 50,000 who captured trout first in this 5% stream, only 5% of them will catch a trout the next time also. Well, 5% of 50,000 is 2,500, so that's the number of fishermen among 1,000,000 fishermen at the 5% stream; only 2,500 of them would catch trout, then another trout. And then, of those 2,500 who caught first a trout, then a trout, only 95% of them will catch a non-trout the next time, because remember, we're assuming at this particular stream that only 5% of the fish are trout, and you're randomly taking them. That means 2,375 fishermen would get a non-trout on the third time—in other words, the hypothesis that these 1,000,000 fishermen are at the stream that contains only 5% trout yields the result that 2,375 fishermen would catch a trout, a trout, and then a non-trout.

All we're doing is looking at this hypothetical world. Now what we're going to do is, of course, do exactly the same thing for all the hypothetical streams. In other words, at the 15% lake, it's going to be a different computation, but the same style, because in the 15% lake, 15% of the fishermen will first catch a trout instead of 5% in the 5% lake. So 15% catch a trout the first time, and then 15% of those catch a trout the next time, and then 85% will catch a non-trout the third time. Doing the arithmetic, we see that 19,125 fishermen will, on average, catch a trout, trout, and non-trout in the 15% lake.

Well, we just repeat this analysis for every hypothetical lake in the entire possibilities of those 10, and in doing so we actually have a chart—yeah, it was pretty boring,—but we have this chart that actually has computed exactly how many of 1,000,000 fishermen in each of the hypothetical 10 universes, how many of them would catch trout, trout, non-trout. And here are those numbers.

Now notice that in the 65% stream—the hypothesized 65% stream, where 65% are trout, 147,875 fishermen of 1,000,000 would have caught trout, trout, and non-trout—there are many more who would have caught it that were in the 65% stream than who would have experienced that particular sequence of fish-catching in the 5% stream.

Well, now our goal is to update our beliefs. The way we do that is we first total up the total number of fishermen among these whole 10,000,000 hypothetical fishermen, how many of them experience the trout, trout, non-trout fish-catching? Well that's just the sum of these numbers, which is 837,500 fishermen among those 10,000,000 caught trout, trout, non-trout. And of those, 2,375 were in the 5% stream, 19,125 were in the 15% stream, and so on.

So for every stream, we can see what fraction of the fishermen who experience trout, trout, and non-trout were in the various streams. For example, there were 2,375 out of 837,500 were in the 5% stream; that's the proportion of the total number of fishermen with that experience who were in the 5% stream. And likewise, 19,125 out of that number were in the 15% stream. That allows us to summarize the new view of the world, because what we did is we were able to then calculate—let's go to this table here—we're able to calculate the probability that, given that we have the evidence of catching two trout and one non-trout, and we did it in all of these streams, we now see that the fraction of the fishermen who caught trout, trout, non-trout in the different streams give us the probability that we were in those different streams.

So here we go. Let me summarize our strategy for how we update our belief system. The first thing is that we started with an a priori assumption, and in this case our a priori assumption before we caught any fish was that each of the 10 hypotheses were equally likely. Then, given that we catch two trout and then one non-trout, we calculated the probability that the hypothesis of being in the 5% stream is correct, and that probability was 2,375 out of 837,500. Then we calculated the probability that the hypothesis of being in the

15% stream was correct, and we computed it—19,125 out 837,500—and so on. This gives us the probability of the different streams—5% stream, 15% stream, 25% stream. These, then, are the probabilities of actually being in the different stream. In other words, we've updated our belief as to what the distribution of possibilities of the world are.

Now we have this new chart that shows that we much more strongly believe that the number of trout in the stream is much more likely to be a 65% stream, or a 55% or a 75% stream than it is to be a 5% stream.

So the summary of the Bayesian procedure, so what we've just demonstrated, is the concept of Bayesian probability. We start with an a priori estimate of the probability of each hypothesized state of the world, we gather information, and then we update our probability estimates, getting an a posteriori estimate. And by the way, I should insert here that Bayesian probability is named after Thomas Bayes, who lived from 1702 to 1761, and he was a Presbyterian minister and a mathematician. He is the one for whom Bayes' theorem that we met in the last lecture is named, and is associated with this strategy of updating our belief system.

Now, suppose that we catch more fish, and I will spare you from going through the details of how we do the updating, but let me just tell you what happens. If we catch now another trout, and then another non-trout, we do almost the same thing, except instead of imagining that we have our 10,000,000 fishermen evenly distributed 1,000,000 fishermen per lake, in order to do the next updating, we imagine that the fishermen are distributed according to our new concept of the probability. In other words, we have in our next view of updating, we have a lot more fishermen who are starting to fish in the 65% stream than are fishing in the 5% stream because our new a priori distribution has now a new bias. It's no longer agnostic. It no longer believes that it's even. We have evidence that changed our sense of the possibilities.

After we apply the same techniques, having caught another trout and non-trout, we have updated our beliefs; and we get a slightly different shaped chart. And so if we summarize our progress, we started with our original agnostic belief, we got evidence, it changed to this shape. We got more evidence, catching another trout and another non-trout, we got this shape, and then in general, if we continue to catch fish, we'll continue to update our probability, and as we catch more fish, the evidence will dominate over our initial estimate of our uncertainty. In other words, our initial guess will play a smaller role as we get more and more evidence. This is reflecting the law of large numbers, that as we get more evidence, it's going to be more compelling than our initial guess was.

And let me just demonstrate this by showing what happens if we catch 13 trout and 7 non-trout. Then the probabilities of being in a 5% stream, or 15% or 25% stream become very low, and most of the probability is centered here. If we continue to catch 26 trout and 14 non-trout—and by the way, we're catching them in proportion of about 65% just to demonstrate the effect on our probability distribution as we catch more trout that are in the probability that is indicating that it's a 65% stream. So after we've caught here 100 fish, 65 of which were trout and 35 were non-trout, then our estimate of the possible percentages of trout in the stream, we have very strong belief that it's a 65% stream, but there's still a 10% chance that it's a 55% stream, and maybe a 6% chance that there's a 75% stream.

Okay. This is the basic strategy of Bayesian probability; the two philosophically different views of probability are the frequentist view, in which probability is defined in terms of a long-run frequency or proportion of outcomes in repeated experiments, versus the Bayesian probability view, which is the view in which probability is interpreted as a measure of the degree of belief, and in this view the concept of probability distribution is applied to a feature of a population to indicate one's belief about possible values of that feature, even though just one of those is actually true.

Well, let's do this same strategy in a medical example, because this is a place where it really comes up in actual practice. Suppose that you go into a doctor's office, and you talk to the doctor, and the doctor has an initial guess about the possibility of the different diseases you may have. It's the example I talked about before: The doctor feels that you have a 50% chance of having a disease A, a 40% chance of having a disease B, and a 10% chance of having a disease C. That's after an initial examination.

Well now, after taking more tests or a more thorough examination, a new symptom is discovered; let's call it S. Well this new symptom has—because of the experience of the doctor, the doctor knows what percentage of people who have disease A exhibit that particular symptom—so this is the conditional probability of S given A. That means that among people who have disease A, what fraction of them have the symptom S? Well, suppose that for disease A only 10% of people who have disease A exhibit that symptom, and 30% of those who have disease B exhibit that symptom, while 80% of those who have disease C exhibit that symptom.

Well, we have more evidence now, and that evidence can be used to update our initial guess of the probability. So what we do is exactly the same thing as with the fish. We say, "Our initial probability was 50%, 40%, 10% probability of disease A, B, and C, respectively." So we imagine 10,000 patients who came to the doctor with the symptoms that you had when you went to the doctor; among those 10,000, 5,000 would have disease A; 4,000 would have disease B; and 1,000 would have disease C.

Now we use the fact of the probability of showing symptom S. Well, 10% of those who are imagined to have disease A would exhibit the symptom S—that's 500. Thirty percent of those who have disease B of the 4,000 who are supposed to have disease B will exhibit the symptom—that's 1,200 patients. And 80% of those who have disease C would exhibit the symptom—that's 800 patients. So of these 10,000 patients who were distributed according to the initial estimate of the probabilities of these diseases A, B, and C, we now see that of those, a total of 2,500 patients will exhibit the symptom S, and they will be distributed according to this fashion. In other words, 20% of those will have, among the 10,000 patients that we imagined to have been distributed according to our a priori estimate of what disease that person had, 20% of the total who also had symptom S would have actually had disease A, 48% would have had disease B, and 32% would have had disease C. What that has done then is it allows us to update our probability of which of the diseases actually is the correct one.

What we have seen, then, is that the concept of having an a priori distribution about what the reality of the world was, and then getting additional evidence, allows us to update our concept of what the possibilities are—and in the case of medical studies and medical diagnosis, this is exactly what we see when we watch the TV and they say, "Let's look at the differential diagnosis." What they're saying is, "If we have a patient who might have disease A, B, or C, and then we do a test that has a new symptom, we say, 'Well, boy, if that patient had disease B, there's less of a chance that it would shown up as it does.'"

So in this medical example, then, we can exhibit it with a new histogram, showing how the old histogram is related. Here's the old histogram, and we see that in the old histogram, A was the dominant possibility, and that would be the disease for which you would be treated. Whereas after finding symptom S, we see that B has now become a 48% chance of being the correct disease, and C in fact now is more probable even than A. This is an example of the philosophy of Bayesian probability.

What I really like about this whole subject is that it captures our concept of how we learn, and how we have an impression of the way the world is, and that by learning, the evidence is used to modify our impression. But then in Bayesian probability we're very specific about saying, "Well, let's actually write down, first of all, what the possible states of the world are that we're considering, and what probabilities we ascribe to each of those different states."

In the next lecture we're going to be talking about a conundrum of probability involving two envelopes that presents sort of a paradox of expected value that brings up intriguing ideas. I'll look forward to seeing you then. Bye for now.

Lecture Twelve
Probability Everywhere

Scope:

One of the strengths of mathematics is its strategy of generalizing and abstracting ideas. In the case of probability, we have mostly considered situations for which a finite number of possible outcomes was possible for a given situation; then, we investigated issues of probability associated with that situation. The techniques we developed can be extended to situations in which infinitely many outcomes are imagined as possible. The two envelopes problem and the St. Petersburg paradox each force us to confront new challenges that arise when infinitely many outcomes are possible.

Probability is a fascinating study that has many real-world applications. Probability presents us with a rich field of intriguing inquiry that contains questions and insights that are mathematical, practical, and philosophical.

Outline

I. Often, mathematical ideas are born by trying to tackle a specific problem. In thinking through how to deal with the specific problem, new ideas are created.

    A. Here is a conundrum known as the two envelopes problem: You are given two envelopes and told that each envelope contains a check for a certain amount of money, and one of the checks is for exactly twice as much money as the other.

    B. You randomly select one of the envelopes and open it. The enclosed check is for a certain amount of money, say d dollars.

    C. Now you can either keep that money, or you can take the contents of the other envelope.

        1. You know that you are as likely to have chosen the lesser amount as you are likely to have chosen the greater amount.

        2. But now you do an expected-value analysis and find a paradoxical situation.

3. There is a 0.5 probability that you have the higher amount and a 0.5 probability that you have the lower amount; thus, the expected value of switching is: $\frac{1}{2}(2d)+\frac{1}{2}\frac{d}{2}=d+\frac{d}{4}=\frac{5}{4}d$. The result is greater than d, so this analysis seems to suggest that you should switch.

4. But this makes no sense, because it is clearly as likely that you have the lower amount as the greater amount. What is wrong?

D. The two envelopes problem brings up a situation we have not dealt with much, namely, one in which the experiment has infinitely many possible outcomes. How can we revise our thinking to cope with infinitely many alternatives?

1. If any amount of money is possible, then there are infinitely many possibilities theoretically.

2. But in reality, huge numbers are not possible. Actually, we have an a priori sense—an expectation—of a probability distribution.

E. We can hearken back to the Bayesian strategy and realize that we have an a priori sense of the probabilities of various amounts.

1. Depending on our a priori beliefs, we are forced to confront reality and realize that we don't have an infinite number of possible amounts of money in the envelopes.

2. We can describe the probabilities by a graph based on the expected-value analysis using the probabilities according to our a priori distribution.

3. The expected-value analysis using the probability distribution that takes into account the infinite number of possible outcomes will give us good guidance about whether to switch envelopes.

F. In addition, we must point out that even when dealing with an infinite number of possible outcomes, we must assign probabilities that total 1.

II. Another famous paradox involved with gambling is the St. Petersburg paradox.

A. Suppose a gambler plays a coin-flipping game, winning $2 for flipping heads. If the gambler flips tails, then heads, he wins \$4. If the gambler flips two tails in a row, then heads, he wins \$8. If the gambler is very lucky, he might flip five tails in a row, followed by heads, to win \$64.

B. How much would you pay to play this game? We can calculate the expected value:

$$\frac{1}{2}2+\frac{1}{4}4+\frac{1}{8}8+\frac{1}{16}16+...= \quad 1+1+1+...=\infty$$

  1. The expected payout is infinity, so it appears you should pay any amount of money to play this game.
  2. The paradox is that you would not make a great deal of money in this game.

C. Simulations show that average payoffs are quite low.

III. Probability predictions are the basis of statistical inference.

A. Statistical inferences boil down to comparing expectations from probability with collected data.

B. If your expectation from probabilistic analysis differs greatly from what you see in the data, you can make the deduction that the concept you had about how the data were being produced must be wrong.

C. Statistics is an important application of probability and is covered in The Teaching Company course *Meaning from Data: Statistics Made Clear.*

IV. Probability is a fascinating field that plays a fundamental role in how we understand our world, from games to science to finance.

A. One recurring theme of the course was that randomness and probability often confront us with situations that are counterintuitive.

  1. Probability offers intriguing and sometimes subtle puzzles, such as the birthday problem, the Monty Hall *Let's Make a Deal*[R] puzzle, and the two-boys puzzle.
  2. All these examples seem wrong, but when our intuition and reality are not in accord, one of them has to give, and it has to be our intuition.
  3. After we have adjusted our understanding to see the truth of these counterintuitive examples, then the

probability results are ones that we can make reliable decisions on.

4. Probability gives us a logically sound way of quantifying uncertainty.

B. Many of the ideas about probability in this course were illustrated in the realm of gambling, because gambling games are fundamentally based on probability.

1. Casinos count on probability to ensure their success.

2. Casinos are the modern world's testament to the Law of Large Numbers.

C. Random behavior that results in regularity in the aggregate is a central feature of our serious, scientific understanding and descriptions of nature.

1. The directions and speeds of molecules are far too numerous to count and describe. Instead, we describe the interactions as the result of a probabilistic description of random motion, with appropriate constraints that describe the molecular behavior.

2. The role of randomness is central to the science of genetics because the whole premise of the subject is that parts of the genetic material from each parent are randomly donated to the offspring.

D. Of course, probability plays a central role in descriptions of our financial world and investments.

1. Investments are viewed as having a probability of rising or falling.

2. Devising an optimal portfolio involves optimizing the probability of success.

V. One of the fundamental sources of our uncertainty about the world is that often, we don't know what is really true among several possibilities.

A. When we sit on a jury, we may not know whether the accused is innocent or guilty. Instead, we have a sense that there is some likelihood of guilt and some likelihood of innocence.

B. As evidence is adduced at the trial, our relative confidence in guilt or innocence shifts.

C. The strategy of Bayesian probability describes the relative strengths of our beliefs and how they are altered by evidence.

VI. Randomness and uncertainty are fundamental parts of reality.

    A. Probability describes what we should expect from randomness.

    B. Probability is a basic tool for making sense of and coping with the reality of randomness and uncertainty in our world.

Readings:

Ivars Peterson, *The Jungles of Randomness: A Mathematical Safari.*

Sheldon Ross, *A First Course in Probability.*

Questions to Consider:

1. Suppose in the St. Petersburg game, the rule was changed so that you received $64 as soon as you flipped five tails in a row and the game then ended. How much should you pay to make it a fair game? Would you play such a game?

2. Some people believe that everything that happens in life happens for a reason. To what extent do you believe that the occurrences of everyday life are random?

Lecture Twelve—Transcript
Probability Everywhere

Welcome back. In this lecture we're going to follow a road that often leads to interesting ideas, and that is the road of trying to understand what appears to be a paradoxical kind of situation; and then in thinking it through, we develop an idea.

In this case our paradoxical situation consists of imagining the following hypothetical kind of a challenge: namely, suppose that you were given the following game to play. You have two envelopes, and you were given these two envelopes, and you shuffle them and choose one at random. And what you're told about these two envelopes is that one of the envelopes contains a certain amount of money, and the other envelope contains twice as much money, exactly twice as much money, but you don't know which envelope contain which, of course. That makes it fun.

Here you have the random envelopes and you close your eyes and you're given an envelope and you open it up and you see how much money is in there; and here we have d dollars—d dollars are in the envelope. So you say, "Well, that's a great idea. That's wonderful to have d dollars," but somebody gives you the option of doing one of two things: Either you can keep the d dollars that you're given in that envelope, or if you would prefer, you could switch to the other envelope. Now, of course, if you hadn't been watching this course you would say to yourself, "What possible difference does it make, because I don't know how much is in either envelope, and so it doesn't matter if I keep the one I have or change?"

But let's see if we can think it through, and we'll discover that we run into sort of a paradoxical kind of analysis. So, having been in this course, you understand that one way to think it through is to imagine an expected value analysis of what would happen if you switched. In other words, you can say, "If I switched, what would be the expected gain, the expected number of dollars I would take home?" Well, let's just go ahead and do this expected analysis and see what happens.

Here's the expected analysis. Now, the envelope that we opened had d dollars. The other envelope has either $(2 \times d)$ dollars or it has $d/2$ dollars, half as much money. What we can do is do an expected value analysis, and here's the way we do expected value—we've done it before—what we do is we say, "Well, since there's an equally likely chance that the other envelope has either $2d$ dollars or it has $d/2$ dollars, the expected value of switching is ($\frac{1}{2} \times 2d + \frac{1}{2} \times d/2$). Let me review this and make sure you understand what this means. This says that half the time you switch you'll get the envelope with the larger amount of money, $2d$ dollars; and half the time you switch you'll get the amount of money with the lesser amount, $d/2$ dollars.

Well, just let's do the arithmetic: ($\frac{1}{2} \times 2d$) dollars is d; ($\frac{1}{2} \times d/2$) is $d/4$; ($d + d/4$) is $5/4(d)$. Now let's make sure that we let this sink in. What this suggests is that on average if you switch, you will make more money than if you stick with the original envelope that you have. Now, I hope that I haven't found a way to have you completely abandon your commonsense. Your commonsense tells you that this is ridiculous. It's ridiculous to think that, getting an envelope with d dollars that it's always better to switch. The reason it's ridiculous is because, suppose you switched. Then you could imagine that that would have been the first envelope you took, and then the same analysis would have told you it would have been better to switch. This is a paradox; this is a paradox What's wrong? There is something wrong.

What is wrong with this kind of analysis? How are we going to deal with it? The problem that's wrong is that we have an infinite number of possible amounts of money in these two envelopes. That's what we're imagining; and we're imagining that the probability of any of those amounts of money is the same. You see, when we did the expected value computation, remember how we did it? We said, "Well, there's a 50% chance that the amount of money in the envelope is $2d$ and a 50% chance that the amount of money in the envelope is $d/2$." But the question is: Is it really possible that every amount of money is equally possible for the two envelopes?

Let's think it through. The answer is no. It is impossible to imagine that the probability of any given amount of money in the envelope is exactly the same as the probability of any other amount of money. And we can think about it in a very practical way. Let's think about this as a real life situation. Suppose somebody literally gave you this envelope and asked you to undertake this experiment. I mean, literally, a person that you know. That is to say, suppose that your neighbor comes over and you discuss this thing, and your neighbor says, "Well, let's just try it, and I'll put real money, or a real check, that will actually be paid, into the envelope." Well, the first thing you realize is that it's not going to be the case that that check will be for, say, $10,000,000. You have an expectation for how much money will actually be in the envelope before you even open any of the envelopes, you see? So the theoretical question of saying, "Well, any amount of money in the envelope is possible," is really not a practical question, because in reality you do, in fact, approach this whole situation with an a priori bias about how much money is possible to be in the envelope.

What we can do is to try to pin down what our bias is. What we're doing is we're harkening back to the Bayesian strategy of having an a priori distribution about our sense of the belief of what's happening in the world. We don't believe that every single possible dollar amount is equally likely. That's not what we really bring to the table, although at first when we first were confronted with this envelope problem, we didn't think about saying, "Well, jeez, I wonder how much is actually going to be in there." That didn't come to our minds, but now that we see the paradox, it's forced us to confront the fact that we must have, inherent in us, an a priori distribution of what we believe the possible values in the envelope might be.

For example, suppose we have this as our distribution of probabilities of how much money is in the envelope. In other words, if we have, for example, let's suppose that we open the envelope and there were $64 in the envelope, and the height above that is indicating what the likelihood is that we assume a priori is the value that's likely to be in the envelope, given the circumstances—who's paying the money, how rich they are, and other circumstances like that—we do have this a priori sense of how much could be in the envelope, and the effect of having that sense is that the two probabilities associated with the probability of the other envelope having twice as much money versus the probability of its having half as much money is that those probabilities are not necessarily equal. And depending on our a priori sense of what's happening, then we don't run into this expected value paradox, because, in fact, by looking at the actual probability, given our a priori sense of the likelihood of twice as much money being in the envelope, then that will tell us whether or not the expected value indicates that it's a good idea to switch.

In this case, for example, where there's $64, since $128 has a smaller probability than the $32, then in this case we might say, "No, it's more likely that the $32 is what's in the other envelope; and consequently I should stick with the amount of money I have." But it's based on using the expected value analysis, which is completely correct, except using not the probability of ½, but of using the different probabilities according to our a priori distribution.

Now, by the way, there's another very basic point that's brought up by this issue of having infinitely many possible values. In other words, even suppose we were just doing an abstract mathematical issue of having a number in the envelope, it's impossible to have a non-zero probability that is equal for infinitely many different things. Because, you know, if you add up the probabilities of each possible outcome, like the probability of having between $1 and $2, and the probability of having between $2 and $3, and between $3 and $4, and $4 and $5, and so on, the sum of those probabilities has to add up to 1. But if they're all the same, you can't have an infinite set of numbers that are all the same and all above 0 that just add up to 1. They would add up to infinity. So any probability distribution of the values in the envelopes has to actually taper off somewhere.

Okay, so this was the two envelopes paradox. I want to tell you another paradox that is involved with gambling that is a famous paradox that arose in the early days of the discussion of probability, and that is the paradox called the St. Petersburg paradox. So this is a game that is played at a casino, and it was imagined to be in St. Petersburg when it was originally proposed. And the game was the following game: You just flip a coin, and if the coin comes up heads you're paid $2. If the coin comes up tails, and then you flip again and get a heads (that is, the first heads is on the second flip), then you get $4. If you shake tails, tails, and then heads, you get $8; three tails then a heads, $16, and so on. You see?

Now, the question comes: What is the value of playing this game? In other words, if somebody were going to give you that payoff—you always make money—how much would you be willing to give them in order to play the game? Well, the way that we have analyzed this before is to do an expected value analysis. If we do it, here's what we find: that with 50% of the time you'll get heads the first time and get $2 as your payoff. A quarter of the time you'll first get a tails, then a heads, and you get a $4 payoff. An eighth of a time you'll get tails, tails, heads, and get an $8 payoff, and so on. Each of these is contributing $1. Therefore, the expected payout is infinity. So from the expected value analysis it would appear as though you should pay any amount of money to play this game—that it's a great thing. The paradox is that of course it would be ridiculous to pay that kind of money because if you do, you'll find that you lose a lot of money. A rational person would not pay very much money to play this game, and in order to demonstrate that we did some simulations.

Here's how the simulations go. We just played the St. Petersburg game, that is, flipping a coin by simulation 1,000 times, and we calculated the average of the payouts per game over those 1,000 times. Do you follow me? In other words, sometimes we got $2, sometimes $4, sometimes $16, and so on, and we took the average of all those different times we played, and we did it 1,000 times and took the average. We did that 10 times, and then we did the same thing for 10,000 trials, and then 100,000 trials. So here are the results of our simulations.

The results are, when we played the game 1,000 times, these are the average payoffs that we got: There was one rather high one, $43, but the other ones were at the most $20. When we played the game 10,000 times, here are the average payoffs: No number was above $22. Now, remember, the expected value that we just computed was infinite, and yet here in reality we're getting average payouts of very small amounts. Here we go; here's where we played the game 100,000 times—simulated, and we simulated 100,000 repetitions, and we did those simulations 10 times. So do you follow me? Ten times we did 100,000 sequences of flips of the coins until we got a head. Look at the payoffs of playing all of these times. Every number here, except for one, number 2, is $22 or less. Number 2, by the way, had a single payoff of $2,097,152. So that was a case where you flipped 20 tails in a row before you got a heads, and paid off that amount of money. Of course, you've got 100,000 attempts, and so these kinds of things will happen—100,000 attempts repeated 10 times.

The St. Petersburg paradox is that although the expected value is enormous, in practice the amount that you would want to pay to play this game is actually rather minimal. One resolution of the paradox is that if there is a maximum amount that the casino has and will ever pay, even if you shook many, many tails in a row, if there's a maximum, then in fact you can compute an expected value that is very modest. For example, if the casino will never pay more than $1,000,000, the expected payout is $21. If it will never pay out more than $1,000,000,000, the expected payout is $31, and if it is something approaching the Gross National Product of the United States (not quite), then you would get an expected payout of about $44. So even though the expected value is enormous, in practice you wouldn't want to pay very much to play this game.

Well, I wanted to now turn to the whole course in general, and say some things in summary about the whole course, and the first thing is about statistics. One thing that this course has not done is to really talk about the major application area of probability, and that is statistics; or maybe not *the* major one, but one of the most common places where probability is applied is in statistics, because probability is the central driving force for making statistical deductions, doing statistical inference. And the reason is that in statistics, the way that statistical inference works is that you see what would be expected from probability, and then you compare it to the data that you actually collect, and if your expectation that you deduce from a probabilistic analysis differs greatly from what you actually see in the data, then you can make the deduction that the concept that you had about how the data were being produced must be wrong. That's the basic concept of statistical inference.

However, we didn't spend time on that because we have another course—the statistics course—that delves into that aspect of the application of probability a great deal, so we just didn't want to repeat it in this course. But I didn't want you to feel that statistics is not an important application of probability.

Let me then conclude the course by making some general remarks, and that is that probability is a fascinating field, and I think we'll all agree that it's fascinating, and that it plays a fundamental role in how we understand our entire world. From games, to science, to finance, we've seen probability apply in all these areas, and so in these last few minutes, I want to look back on the whole course, and somehow point out some of what I see as the overall effect of what we've seen.

And to me, the bottom line is that randomness and uncertainty are fundamental parts of our experience of life, and the role probability plays is to describe in quantitative detail what we can expect from that randomness and that uncertainty. Well, one of the recurring themes that we saw throughout the course—in fact, even including in today's lecture about the two envelopes and the St. Petersburg paradox—one of the recurring themes was that randomness and probability often confront us with situations that are very counterintuitive.

For example, we saw that in 97% of the cases, there's a 97% probability, that if you're in a room with 50 random people, then at least two of them will share the same birth date. We saw that it was better to switch. If we ever get on a revival of the *Let's Make a Deal*® show, let's make sure that we switch. And we saw these really bizarre things, like the bizarre influence on the probability of having two boys if a stranger comes up and has two children, and they say, "Oh, well one of them was a boy born on a Tuesday," that somehow knowing that it was born on a Tuesday changed the probability that that person has two boys as their children—that was bizarre.

And probably all of those examples probably all seem simply wrong at first, but then the question that I want to bring up now is: What should we make of these kinds of examples? And the answer is this, that when our intuition doesn't accord with reality, then one of those two things, our intuition or reality, has to give, and what has to give is our intuition.

You see, in all of these cases, the probabilistic analyses that we did were correct, and they really do predict what actually would happen in the scenarios that we described. So when we are confronted with a situation where our intuition is jarred, it's because our intuition needs to be retrained, and that's the purpose of these really, what I view as wonderfully thought-provoking conundrums. They challenge us. They challenge us to rethink the things that are actually mistaken biases. Our intuition is mistaken.

But when we come to understand the correct descriptions of the results and we really understand the analysis, then what we should do is to act according to this more accurate view of the world. After we've adjusted our understanding to see the truth of even these things that originally appeared to be completely counterintuitive, these examples that are so counterintuitive, but after we come to understand them, then the probability results are ones that we can actually make reliable decisions on. It literally is really true that if you are in a room with 50 random people, it's almost certain that at least two of them will have the same birth date, and you can make bets on it and win bets. It really is descriptive of the way the world actually operates.

Well, probability gives us a practical and a logically sound way of quantifying uncertainty, and it's up to us to understand and to accept its results, and to realize that these probabilistic descriptions of reality really do meaningfully give us guidance about what to expect.

Many of the ideas that we introduced about probability were illustrated in the realm of gambling, and that's because gambling games are fundamentally based on probability. Casinos count on probability to ensure their success, and casinos can reliably count on profits based on expected outcomes of random trials. There are conspicuous examples of the confidence that we can place on the fact that random events, even though they're not determined individually but in the aggregate, random events, if they're repeated many, many times, will display a regularity that can be completely relied upon. In other words, casinos are the modern world's testament to the law of large numbers; you'll recall the law of large numbers states that if you repeat a trial that has a random outcome many, many times, the experiments will exhibit the predictions of probability with ever increasing accuracy.

But the fact that random behavior results in regularity in the aggregate isn't just a central feature of games of chance and gambling and casinos, but it really is the center feature of our serious scientific understanding, and our descriptions of nature. The scientific view is that we view every breath of air we take as being composed of countless molecules moving about in ways that we couldn't hope to describe molecule by molecule.

You know if we think about the second law of thermodynamics: We imagine a box that has hot molecules on one side and cold ones on the other, and the hot ones are faster moving, and the cold ones are slower moving on average, and then we remove the partition, and we're trying to describe what happens. Well, we don't describe it molecule by molecule. The directions and the speeds of the molecules, they're far too numerous to count and describe; we couldn't hope to do it that way. Instead, we just abandon that idea of describing what happens as a deterministic system determined by the speeds and directions of each individual molecule, even though that's actually what's happening. But instead we describe the interactions as what will be expected as a result of a probabilistic description of random motion—and where of course we put the appropriate constraints that describe the molecular behavior, on what the distribution of fast molecules and slow molecules is.

Now, there is a different branch of chemistry and physics that actually describes how single molecules interact with other single molecules, but if we want to know what happens if we stir a liquid into another liquid, the result is described by probabilities.

Well, in no scientific arena is the role of probability more central than in the science of genetics; and that's because the whole premise of the subject is that parts of the genetic material from each parent are randomly donated to the offspring. The challenge of genetics is to understand what we can expect from random combinations of that sort. Well, the results are challenging. They're challenging to understand because, although they have probabilistic regularity, such as for example we saw expressed in the Hardy–Weinberg equilibrium theorem. It tells us that we can expect a certain kind of stability on average, and yet the random walk of genetic drift tells us that we have to expect deviation from that stability by randomness alone. But when we learn that, we shouldn't throw up our hands and say, "Oh, so randomness and probability tell us nothing, because anything might happen." No, that's just not true.

On the contrary, our knowledge of probability lets us understand and put into perspective the challenging blend of probabilistic regularity along with the deviation from that regularity, and that deviation occurs at an expected frequency and at an expected rate. Just like we can't know when we throw a die whether it will come up a 3 or not, but there's a specific meaning to say that there's only a 1/6 probability of it coming up a 3, and a 5/6 probability of its not coming up a 3, and that insight is something that we can rely upon.

Probability gives us definite insights. It gives us insights into what will happen over generations to populations' characteristics. And probability can be used, by the way, to model all sorts of other things in the biological realm; such as, for example, the spread of diseases, is an example.

Well, of course, it's expected that probability would play a central role in descriptions of our financial world and investments, because investments are viewed as having a probability of rising or a probability of falling; devising an optimal portfolio involves the optimization of some measure of the probability of its being successful. But unfortunately that entails the uncomfortable reality that truly excellent decisions may, by chance alone, have unfortunate outcomes, and so recognizing the role of randomness and probability in our everyday lives puts a really challenging perspective on how we might react to successes and failures that entail a probabilistic component. You know, good decisions can have bad outcomes by chance alone; and that's one, I think, of the difficult parts of the nature of probability. It's one of its greatest challenges, philosophically.

Well, one of the most fundamental sources of our uncertainty about the world is that often we don't know what's really true among several possibilities. For example, if we sit on a jury we don't know whether the criminal in front of us is innocent or guilty, but instead we have a sense that there's some likelihood that the person is guilty and some likelihood that the person is innocent, and as evidence is adduced at the trial, our relative confidence in the guilt or innocence of that person shifts, and the strategy that Bayesian probability describes is that it looks at the relative strengths of our beliefs and it sees how those strengths are altered by evidence. It models a basic experience of life; namely, what uncertainty really is. It talks about what that is; namely, you're giving a probability to different possible states of the world, and it shows how our beliefs change when we get more evidence.

To me, the cumulative effect of the whole course is to underscore the reality that randomness and uncertainty are fundamental parts of reality. Probability describes what we should expect from randomness, and consequently probability is a basic tool for making sense of and coping with the reality of randomness and uncertainty. By its nature, probability forces us to confront directly two very different fundamental views of the world. I want to show you two quotes, thinking there's a good probability, a good chance that you'll like one of them—and both of them are from ancient Greece so in fact neither of the speakers reflects the modern centrality of probabilistic thinking in the descriptions of our world, but they do capture a philosophical dichotomy about the nature of reality. So here are the two quotes.

Leucippus said, "Nothing occurs at random, but everything for a reason and by necessity."

Democritus: "Everything existing in the universe is the fruit of chance."

So to end this course, I think we should once more put probability to the test, and see if it works. So let's once again take our 60 dice, roll them, and see if we can have probability randomly show us some regularity. Here we go.

Thanks for watching. Bye for now.

Timeline

1733 ....Abraham de Moivre publishes an account of the normal approximation for the binomial distribution for a large number of trials. This improves upon Jacob Bernoulli's Law of Large Numbers. This account will be included in the 1756 edition of De Moivre's *The Doctrine of Chances*, a treatise on probability first published in 1718.

1738 ....Daniel Bernoulli publishes *Exposition of a New Theory on the Measurement of Risk*, an early look at probability theory and economic decision making.

1820 ....Pierre-Simon Marquis de Laplace publishes a seminal work on probability.

1827 ....Robert Brown, a botanist, while observing the motion of pollen grains, hypothesizes underlying mechanics for erratic movements. This later led Bachelier and Einstein to study and make rigorous Brown's work. The mechanics are now known as Brownian motion in his honor.

1837 ....Simeon Denis Poisson publishes *Recherches sur la probabilité des jugements en matière criminelle et matière civile*, which introduces the expression *Law of Large Numbers* and in which the Poisson distribution first appears.

1853 ....Augustin-Louis Cauchy presents an outline of the first rigorous proof of the central limit theorem, which is a generalization of the Law of Large Numbers.

1867 ....Pafnutii Lvovich Chebyshev publishes a paper, *On Mean Values*, which uses Irenée-Jules Bienaymé's inequality to give a generalized Law of Large Numbers.

1887 ....Pafnutii Lvovich Chebyshev publishes *On Two Theorems*, which gives the basis for applying the theory of probability to statistical data, generalizing the central limit theorem of de Moivre and Laplace.

1900 ....Louis Bachelier publishes the first mathematical approach to Brownian motion in his Ph.D. thesis, *Théorie de la Spéculation*.

1905 ....Einstein publishes three groundbreaking scientific papers. The third and least famous of the three (the first won the Nobel Prize for Physics and the second was on special relativity) detailed a mathematical treatment of Brownian motion.

1919 ....Paul Levy delivers three lectures at the École Polytechnique, highlighting entirely new areas of research in probability theory.

1938 ....Kolmogorov publishes the influential *Analytic Methods in Probability Theory*.

1942 ....Kiyosi Ito publishes *On Stochastic Processes (Infinitely Divisible Laws of Probability)*, a groundbreaking paper.

1944	Von Neumann and Morgenstern publish *Theory of Games and Economic Behavior*, the first text on the new field of game theory.
1950	William Feller writes the first volume of his famous *Introduction to Probability Theory and Applications*.
1953	Joseph Leo Doob publishes *Stochastic Processes*, a now classic text on stochastic (probabilistic) analysis and martingale theory.
1966	Norbert Wiener publishes *Nonlinear Problems in Random Theory*.
1966	MIT mathematician Ed Thorp publishes *Beat the Dealer*, a popular work on applying probabilistic thinking in the game of blackjack in Las Vegas casinos.
1969	Fischer Black and Myron Scholes write their seminal paper on a mathematical and probabilistic approach to pricing options.
1973	Robert C. Merton publishes *Theory of Rational Option Pricing*.
1994–1998	Long-Term Capital Management experiences its strong profitable run, then collapses.
1997	Robert Merton and Myron Scholes, applied mathematicians, win the Nobel Prize for Economics for their work in options-pricing theory.

Glossary

Bayesian probability: The view in which probability is interpreted as a measure of degree of belief. In this view, the concept of probability distribution is applied to a feature of a population to indicate one's belief about possible values of that feature. The principal result of experiments or more evidence is to update such a probability distribution, indicating a change in belief. The Bayesian viewpoint is in contrast to the frequentist view.

Bayes' theorem: A mathematical equation relating two conditional probabilities: $P[A|B] = \dfrac{P[B|A]P[A]}{P[B]}$.

Bell's theorem: A theorem asserting that a particular inequality of certain probabilities would be true if intuitive concepts of local realism were true of particle physics. The theory of quantum physics violates the inequality. Quantum theory implies that when one particle of an entangled pair of particles is observed, the other particle in the pair, which could be distant, instantaneously undergoes a state change. Bell's theorem implies that this aspect of quantum theory cannot be explained by hidden local variables.

chance: An informal term that tries to capture the same notion as the term *probability*.

choose function: The mathematical function that computes the number of ways n distinct objects can be used to form a group of k objects. This value is called "n choose k," or "the combination of n taken k at a time." It is sometimes written $\dbinom{n}{k}$. Its value is $\dfrac{n \cdot (n-1) \cdot (n-2) \cdot \ldots \cdot (n-k+1)}{k \cdot (k-1) \cdot (k-2) \cdot \ldots \cdot 2 \cdot 1} = \dfrac{n!}{k!(n-k)!}$ (see **factorial**). For example, 5 choose 3 equals 10.

complementary event: The event complementary to a given event is the set of all possible outcomes that are not in (or do not satisfy or do not represent) the given event. For example, in rolling two dice, one event is: "The sum of the dice is 8." Its complementary event is: "The sum of the dice is not 8."

conditional probability: The probability of an event under the assumption of the existence (or happening or satisfaction) of another event. For example, in rolling a blue fair die and a red fair die, the conditional probability of the event "the sum of the dice is 8," given the event "the blue die is 3 or 6," is $\frac{2}{12} = \frac{1}{6}$, because there are 12 possible outcomes with the blue die being 3 or 6, and two of those (blue 3 and red 5; blue 6 and red 2) sum to 8. We would say, "The probability of the sum being 8 given that the blue die is 3 or 6 is $\frac{1}{6}$."

deterministic model: A mathematical description of a phenomenon or mechanism that does not depend on randomness. Every time the model is executed with the same initial conditions, the result (prediction) will be the same. Contrast with **probabilistic model**.

disjoint events: Two (or more) events that cannot both happen (for one experiment). Each possible outcome of the experiment is in (or satisfies or represents), at most, one of the events. For example, in rolling two dice, the event "the sum is 8" is disjoint from the event "there is a 1."

event: A set of possible outcomes of an experiment, trial, or observation. For example, for the trial of rolling a blue die and a red die, a possible event is: "The sum of the dice is 8." This event consists of the following five outcomes: blue 2 and red 6, blue 3 and red 5, blue 4 and red 4, blue 5 and red 3, blue 6 and red 2. Compare to **outcome**.

expected value: Assuming a numerical value is associated with each possible outcome of an experiment (or a trial or an observation), the expected value of the experiment is the weighted average of the values, where each weight is the probability of the associated outcome. The expected value is a number that summarizes the possible values. The term can be misleading, because often the expected value as a number is not associated with any possible outcome. For example, in the experiment of flipping a fair coin, if the value 2 is associated with heads and the value 5 with tails, then the expected value is 3.5 (which is neither 2 nor 5 and, hence, hardly to be "expected"). More formally, it is the expected value of a random variable that is defined, rather than the expected value of an experiment.

factorial: The mathematical function that computes $n(n-1)(n-2)(n-3)...(3)(2)(1)$. This value is denoted $n!$ and is read, "n factorial." It represents the number of orderings in which n distinct objects can be listed.

fair: When used in such phrases as "a fair coin" or "a fair die," this term indicates the ideal situation in which the probability of any of the possible outcomes is the same.

flush: In poker, a hand of five cards in which all the cards are of the same suit but cannot be placed in sequential order. See **straight** for examples of cards in sequential order. Compare to **straight flush**.

frequentist probability: The view in which probability is defined in terms of long-run frequency or proportion in outcomes of repeated experiments. This concept of probability is applied to outcomes of actual or hypothetical experiments that have an element of randomness. But in the frequentist view, probability is not used as a measure of knowledge or belief of the possible values of a quantity that does not have a random element. The frequentist viewpoint is in contrast to the Bayesian view.

independent events: Two events are independent if one event's occurring does not affect the probability that the other occurs. If A and B are independent events, then $P(AB) = P(A)P(B)$; that is, the probability that both A and B occur is the product of the probabilities that each occurs. For example, in flipping two coins, assuming that the results of one flip don't affect the results of the other, then the probability of both coins landing on heads is the product of the probability that the first coin lands on heads times the probability that the second coin lands on heads.

Law of Large Numbers: The theorem that the ratio of successes to trials in a random process will converge to the probability of success as increasingly many trials are undertaken.

mutation: A change in a gene of an organism. Some mutations are inherited by offspring of the organism that suffered the mutation. Mutations are often modeled as occurring randomly. Probabilistic models make assumptions on the rate of mutations that are passed to offspring. From these models, conclusions are drawn about the evolutionary history of species.

odds: An alternative way of expressing the probability of an event by stating the ratio: the probability that the event happens divided by the probability that the event does not happen. For example, if the probability of an event is 20%, the odds are $\frac{20}{80}$, or $\frac{1}{4}$. This is sometimes stated, "four to one against."

option: In the financial markets, a contract giving the holder the right to buy a prescribed asset (such as a certain number of shares of a specific stock) at a prescribed time in the future for a prescribed amount of money, or a contract giving the holder the right to sell a prescribed asset at a prescribed time in the future for a prescribed amount of money, or other related contracts.

outcome: A possible specific result of an experiment, trial, or observation. For example, for the trial of rolling a blue die and a red die, one possible outcome is blue 3 and red 5. Compare to **event**.

permutation: An ordering of distinct objects. For example, there are 24 permutations of the four cards ace of spades, king of diamonds, queen of diamonds, and eight of hearts because there are 24 different ways to order those four cards.

poker: A card game (with several variations) played with an ordinary deck of 52 cards, in which five-card sets are compared to see which is "better." The ordering is based on the probabilities of various possible features of a five-card set; rarer features win.

prime number: A whole number (an integer) bigger than 1 that is not evenly divisible by any positive whole number except itself and 1.

probabilistic model: A mathematical description, with random aspects, of a phenomenon or mechanism. The model could consist of mathematical formulas that refer to random numbers. Thus, one execution of the model will generally give different results than another execution. Contrast with **deterministic model**.

probability: As the term is used in mathematics, a number between 0 and 1 (or 0% and 100%) applied to a possible future event that quantifies the likelihood of the event's occurring, or that number applied to a statement that quantifies our degree of belief in the truth of the statement.

probability distribution: A discrete probability distribution is a table, function, or graph that assigns a probability to each possible outcome. For the continuous case, in which any real value is a possible outcome, the probability distribution can be viewed as a graphed curve that has an area of 1 under the curve and above the horizontal x axis. The probability of an outcome being between value a and b is equal to the area under the part of the curve between $x = a$ and $x = b$.

random variable: The assignment of a number to each possible outcome of an experiment. The term *random variable* is an unusually poorly chosen term, because it denotes something that is neither random nor a variable. We avoided using this term in this course.

random walk: A sequence of positions of an object that takes one step each second (or other unit of time), in which the direction of each step is random. The direction of each step is randomly chosen independent of any other step. An example of a one-dimensional random walk is formed by flipping a coin to determine whether the next step should be forward or backward.

randomness: The aspect of life, or a system, or a pattern, or a mathematical model that is unpredictable even in theory or unpredictable because of lack of detailed knowledge. Randomness in a system implies that the behavior of the system can be different even if the system is subjected to identical circumstances. Although random occurrences are not predictable, they exhibit regularity in the aggregate after many repetitions.

roulette: A gambling game in which a small ball settles into one of 38 slots in a wheel as the wheel is spun and slows. The slots are numbered 0, 00, 1, 2, ... , 36. Presumably, each slot is equally likely on any given spin of the wheel to be the stopping point for the ball. Note: European roulette wheels have only 37 slots (no 00).

stochastic model: Synonym for **probabilistic model**.

straight: In poker, a hand of five cards that can be put in sequential order, with not all five cards being of the same suit. Examples include ace, 2, 3, 4, 5; 9, 10, jack, queen, king; and 10, jack, queen, king, ace; but not jack, queen, king, ace, 2. Compare to **straight flush**.

straight flush: In poker, a hand of five cards that can be put in sequential order and in which all five cards are of the same suit. See **straight** for examples of sequential order.

uniform distribution: A probability distribution in which every possible value is equally likely.

weighted average: Given a set of numbers $\{a, b, c, d, ...\}$ (thought of as values of some quantity) and a weight for each number $(w_a, w_b, w_c, w_d, ...)$, the weighted average is the value $aw_a + bw_b + cw_c + dw_d + ...$. The weights must add up to 1 and must be non-negative.

Biographical Notes

Bayes, Thomas (1702–1761). British nonconformist minister. Little is known about Bayes's life, save that he was educated at Edinburgh University and was a member of the Royal Society. His major contribution to the field of probability was the work he did on the inverse probability problem. At the time, the calculation of the probability of a number of successes out of a given number of trials of a binomial event was well known. Bayes worked on the problem of estimating the probability of the individual outcome from a sample of outcomes and discovered the theorem for such a calculation that now bears his name.

Bernoulli, Jacques (often called Jacob or James, 1654–1705). Professor of mathematics at Basel and a student of Leibniz. He formulated the Law of Large Numbers in probability theory and wrote an influential treatise on the subject.

Black, Fischer (1938–1995). Applied mathematician and economist. Worked both in academia and on Wall Street. Pioneer in the field of options pricing and among the first to bring higher mathematics to the financial sector. Held long-standing beliefs about the inherent uncertainties in the markets. Most famous for coauthoring the Black-Scholes formula, for which his coauthor, Myron Scholes, received the Nobel Prize in 1997.

Cardano, Girolamo (1501–1576). Italian mathematician. An avid gambler, he was the first to explore the mathematics of probability in order to improve his game play. He also recorded the first calculations with imaginary numbers. Cardano was the first to understand that there are fundamental scientific and mathematical principles guiding events previously only describable by chance.

Cauchy, Augustin-Louis (1789–1857). French mathematician and engineer. Professor in the École Polytechnique and professor of mathematical physics at Turin. He worked in number theory, algebra, astronomy, mechanics, optics, and analysis. His contribution to probability and statistics was the production of the outline of the first rigorous proof of the central limit theorem, which is a generalization of the Law of Large Numbers.

Chebyshev, Pafnutii Lvovich (1821–1894). Russian mathematician, founder of the St. Petersburg School of Mathematics. He made fundamental contributions to the theory of probability and statistics, including generalizations of the central limit theorem, which is itself a generalization of the Law of Large Numbers.

de Moivre, Abraham (1667–1754). French-English mathematician. Born in France and educated at the Sorbonne in mathematics and physics, de Moivre, a Protestant, emigrated to London in 1688 to avoid further religious persecution. A future fellow of the Royal Society of London, de Moivre supported himself in England as a traveling mathematics teacher and by selling advice in coffee houses to gamblers, underwriters, and annuity brokers. De Moivre is recognized in statistics as the first to publish an account of the normal approximation to the binomial distribution. In fact, some of de Moivre's methods are so ingenious as to be shorter than modern demonstrations of solutions to the same problems.

Doob, Joseph Leo (1910–2004). American mathematician. Produced substantial work on probability theory, stochastic processes, potential theory, and much more. Also authored several seminal texts on probability theory.

Einstein, Albert (1879–1955). Probably the most famous scientist of all time. In addition to his well-known work in several areas of physics, in 1905, he presented one of the first mathematical treatments of Brownian motion. It was Einstein's interest in statistical mechanics that led him to explore Brownian motion.

Fermat, Pierre de (1601–1665). French lawyer and mathematician. Through an interest in games of chance, Fermat used his mathematical prowess to study the mathematics of chance. Following a brief correspondence with Pascal, the two came to be considered joint founders of mathematical probability.

Huygens, Christiaan (1629–1695). Dutch astronomer and mathematician. While most famous for his discoveries about the planet Saturn and his invention of the pendulum clock, Huygens was also an early pioneer of the mathematics of probability. Following a meeting with Fermat, he presented the first printed work on probability theory.

Ito, Kiyosi (b. 1915). Japanese mathematician and statistician. His contribution to probability theory was to develop the notion of stochastic (probabilistic) differential equations.

Kolmogorov, Andrei Nikolaevich (1903–1987). Russian mathematician who ranks among the greatest of the 20th century. A formalist who helped axiomatize probability.

Laplace, Pierre-Simon Marquis de (1749–1827). French mathematician and astronomer. Professor at the École Normale and École Polytechnique, known for his contributions to calculus, analysis, probability theory, and physics. One of the earliest mathematicians to formalize the theory of probability.

Levy, Paul Pierre (1886–1971). French mathematician. A pioneer in modern probability theory. Not a formalist like his contemporary, Kolmogorov; an important class of stochastic processes bears his name.

Markov, Andre Andreevich (1856–1922). Russian mathematician. Member of the St. Petersburg Academy of Science. Markov worked on the Law of Large Numbers and random walks.

Merton, Robert Carhart (b. 1944). Applied mathematician. Student of Nobel laureate Paul Samuelson. Credited with being among the first to bring stochastic calculus and other sophisticated probabilistic tools to finance. Helped develop the Black-Scholes pricing formula (also called Merton-Black-Scholes). He developed probabilistic and analytic theorems that paved the way for the now-high-profile field of financial engineering. Recipient of the 1997 Nobel Memorial Prize in Economics.

Neumann, John von (1903–1957). Hungarian mathematician and one of the original members of the Institute of Advanced Study at Princeton University (along with Albert Einstein). A genius who contributed to many areas of mathematics and physics, he is most popularly known as the inventor of game theory. He authored a celebrated text, *Theory of Games and Economic Behavior*.

Newton, Sir Isaac (1642–1727). English mathematician and scientist known for the discovery of the law of gravity and as one of the fathers of calculus. Within the field of probability, he is known for his proof of the binomial theorem. There is also evidence that he gave thought to the variability of the sample mean, the basis for the central limit theorem. In his last work, *The Chronology of Ancient Kingdoms Amended*, published posthumously in 1728, Newton estimated the mean length of a king's reign to be between 18 and 20 years.

Pascal, Blaise (1623–1662). French mathematician and philosopher. In the summer of 1654, he exchanged a series of five letters with Fermat, in which they explored a dice game. The first question they considered was how many times one must throw a pair of dice before one expects a double six, as well as how to divide the stakes if a game is incomplete. Because of this correspondence, they are usually considered the cofounders of probability.

Poisson, Simeon Denis (1781–1840). French mathematician. He published *Recherches sur la probabilité des jugements en matière criminelle et matière civile* in 1837, marking the first appearance of the Poisson distribution, originally found by de Moivre, which describes the probability that a random event will occur in a time or space interval under the conditions that the probability of the event's occurring is very small. Poisson also introduced the expression *Law of Large Numbers*, by which he meant that, for a larger number of trials, the proportion of successful outcomes exhibits statistical regularity. Although we now rate this work as of great importance, it found little favor at the time, the exception being in Russia, where Chebyshev developed his ideas.

Scholes, Myron (b. 1941). Applied mathematician and economist. Coauthor of the Black-Scholes options-pricing formula. Recipient of the 1997 Nobel Prize in Economics. Scholes laid down fundamental mathematical assumptions that still dominate derivatives pricing in the financial markets today. He was a partner at the famously ill-fated hedge fund Long-Term Capital Management.

Wiener, Norbert (1894–1964). Applied mathematician. He mathematically extended the work done by Einstein on Brownian motion (hence, the results are often called Wiener processes). In addition, he generalized and abstracted several fundamental notions and definitions in probability theory, laying the foundation for Ito's work on stochastic analysis.

Bibliography

Albert, Jim, and Jay Bennett. *Curve Ball: Baseball, Statistics, and the Role of Chance in the Game.* New York: Copernicus Books, 2003. Introduces the fundamental concepts of statistics through applications to historical baseball. Tackles in detail such questions as the best hitter and hitting streaks.

Berry, Donald A. *Statistics: A Bayesian Perspective.* Belmont, CA: Duxbury Press at Wadsworth Publishing Co., 1996. An excellent elementary introduction to statistics that includes some probability, with many interesting real examples, most from medicine. The Bayesian approach is used.

Burger, Edward B., and Michael Starbird. *Coincidences, Chaos, and All That Math Jazz: Making Light of Weighty Ideas.* New York: W.W. Norton, 2005. A lighthearted but authentic presentation of some intriguing ideas of mathematics, including coincidences and other probabilistic phenomena.

———. *The Heart of Mathematics: An invitation to effective thinking*, 2nd ed. Emeryville, CA: Key College Publishing, 2005. This award-winning book presents deep and fascinating mathematical ideas in a lively, accessible, and readable way. The review in the June-July 2001 issue of the *American Mathematical Monthly* said, "This is very possibly the best 'mathematics for the non-mathematician' book that I have seen—and that includes popular (non-textbook) books that one would find in a general bookstore."

Charlesworth, Brian, and Deborah Charlesworth. *Evolution: A Very Short Introduction.* New York: Oxford University Press, 2003. This is one of the books in the "Very Short Introduction" series of the Oxford University Press. It is a brief but thorough account of the current status of evolution. It includes discussion of mutations, natural selection, and genetic drift.

Derman, Emanuel. *My Life as a Quant.* Hoboken, NJ: John Wiley and Sons, 2004. An autobiography of a successful probabilist working on Wall Street.

Feller, William. *An Introduction to Probability Theory and Its Applications.* Hoboken, NJ: John Wiley and Sons, 1968. Technical but classic text on probability.

Gordon, Hugh. *Discrete Probability.* New York: Springer, 1997. A mathematically rigorous book that starts from the beginning. Treats such topics as how to count combinatorially and has many exercises. Sticks to mathematical examples rather than applications.

Hacking, Ian. *The Emergence of Probability.* Cambridge, UK: Cambridge University Press, 1975. A history of probability, emphasizing the philosophical transitions and early applications that represent breakthroughs in how our world is perceived.

————. *The Taming of Chance.* New York: Cambridge University Press, 1990. Explores the changes in how chance, probability, and statistics were perceived and employed during the 18th, 19th, and 20th centuries.

Haigh, John. *Taking Chances; Winning with Probability.* New York: Oxford University Press, 2003. Illustrates ideas and results of probability using lotteries and games of all sorts, including casino games, dice, football pools, and sports. Includes discussion of the non-intuitive behavior of ties in coin flipping and random walks.

Heyde, C. C., and E. Seneta, eds. *Statisticians of the Centuries.* New York: Springer-Verlag New York, 2001. This book contains short biographies of statisticians from the 16th to the 20th centuries, many of whom made important contributions to probability and its uses.

Huff, Darrell. *How to Lie with Statistics,* New York: W.W. Norton, 1954. This charming little book has been in continuous publication since 1954. Although it is more about statistics than probability, it is eminently readable and cheerfully describes methods to mislead with statistics.

Jaynes, E. T. *Probability Theory: The Logic of Science.* Edited by G. Larry Bretthorst. Cambridge, UK: Cambridge University Press, 2003. This book makes the case for the Bayesian approach to probability and statistics, pointing out the difficulties in the orthodox approach. The book is technical.

Johnson, Norman L., and Samuel Kotz, eds. *Leading Personalities in Statistical Sciences: From the Seventeenth Century to the Present.* New York: John Wiley and Sons, 1997. Presents biographies of more than 100 statisticians and probabilists of the last four centuries.

Laplace, Pierre-Simon Marquis de. *A Philosophical Essay on Probabilities.* Mineola, NY: Dover Publications, 1951 (actually written about 1820). An early and famous work on inductive reasoning, using probability in what we now would call the Bayesian style. The terminology and wording is very old-fashioned and non-standard by today's conventions. Laplace wrestles with fundamental questions about making conclusions based on partial evidence. The most famous section is the computation of the probability that the Sun will rise tomorrow.

Levinson, Horace C. *Chance, Luck and Statistics*. Mineola, NY: Dover Publications, 1963. A down-to-earth, readable, elementary treatment of basic ideas and computations of probability, using mostly concrete settings, often gambling games such as poker as examples.

Lewis, Michael. *Moneyball: The Art of Winning an Unfair Game*, New York: W.W. Norton, 2003. A delightful story of applying probability and statistics in baseball management.

Lowenstein, Roger. *When Genius Failed*, New York: Random House, 2000. The story of the Nobel laureates' foray onto Wall Street.

Moore, Peter G. *The Business of Risk*. New York: Cambridge University Press 1983. Applications of probability to actual business situations. Gives a user's point of view of probability rather than a mathematician's. Has the flavor of a fairly sophisticated economics book, with some formulas and graphs that are essential to the content.

Morgenstern, Oskar, and John von Neumann. *Theory of Games and Economic Behavior*, commemorative edition. Princeton, NJ: Princeton University Press, 2004. The famous book from the inventors of the subject.

Mosteller, Frederick. *Fifty Challenging Problems in Probability with Solutions*. Mineola, NY: Dover Publications, 1965. A nice collection of probability problems, complete with carefully worded solutions.

Paulos, John A. *A Mathematician Plays the Stock Market*. New York: Basic Books, 2003. Engaging, witty stories about what mathematical thinking can disclose about the stock market.

Peterson, Ivars. *The Jungles of Randomness: A Mathematical Safari*. New York: John Wiley and Sons, 1998. Ivars Peterson is a popular author of general mathematics books for the public. This book discusses the subtleties of the concept of randomness.

Rapoport, Anatol. *Two-Person Game Theory*. Mineola, NY: Dover Publications, 1966. A readable classic on game theory. Includes treatment of cases in which the best strategy for a player is a probabilistic one.

Rosenthal, Jeffrey S. *Struck by Lightning: The Curious World of Probabilities*. Washington, DC: Joseph Henry Press, 2006. This book explains the ideas of probability and its appearance in everyday life.

Ross, Sheldon. *A First Course in Probability*. Upper Saddle River, NJ: Prentice Hall, 2001. Good introductory text to college-level probability theory.

Salsburg, David. *The Lady Tasting Tea: How Statistics Revolutionized Science in the Twentieth Century*. New York: Henry Holt & Co., 2001. Presents the history of statistics during the 20th century with humor and insight. Historical anecdotes bring the topics of statistics to life. This book is very readable and enjoyable, though mostly about statistics.

Stewart, Ian. *Does God Play Dice? The New Mathematics of Chaos*. Malden, MA: Blackwell Publishing, 2002. Explores the topics of mathematical chaos and randomness in the world.

Taleb, Nassim N. *Fooled by Randomness*. New York: Random House, 2004. A wonderful book on the hidden nature of chance and luck in our lives.

Thorp, Ed. *Beat the Dealer: A Winning Strategy for the Game of Twenty-One*. New York: Vintage, 1966. The original MIT-versus-Las Vegas book. Genuine mathematical strategies that casinos would never actually let you use.

Yaglom, A. M., and I. M Yaglom. *Challenging Mathematical Problems with Elementary Solutions*, Volume 1. Mineola, NY: Dover Publications, 1964. Contains challenging and fairly sophisticated probability and other mathematical puzzles. The solutions often involve clever ways of looking at problems.

Internet Resources:

Chance Magazine. www.amstat.org/publications/chance/index.html. Accessible articles about statistics and probability and their applications.

Index of Biographies. School of Mathematics and Statistics, University of St. Andrews, Scotland. www-groups.dcs.st-andrews.ac.uk/~history/BiogIndex.html. This website gives biographical information about thousands of noted mathematicians. Both chronological and alphabetical indexes are presented, as well as such categories as female mathematicians, famous curves, history topics, and so forth.

Probability. Cut-the-Knot. www.cut-the-knot.org/probability.shtml. This site includes a list of delightful probability puzzles.

The R Project for Statistical Computing. Department of Statistics and Mathematics, Vienna University of Economics and Business Administration. www.R-project.org. R is a language and environment for statistical computing. This website contains a downloadable statistical and probabilistic software tool for computing and graphing statistical probabilistic calculations.

Wizard of Odds. www.wizardofodds.com. Excellent site run by a mathematician who works in Las Vegas. He includes probabilistic analysis of hundreds of different gambling games and scenarios.